Rita Sundari

A Brief Review on Urban Dynamics in Rapid Growing Cities

AF166891

Rita Sundari

A Brief Review on Urban Dynamics in Rapid Growing Cities

Urban development: prospects and constraints

LAP LAMBERT Academic Publishing

Impressum / Imprint
Bibliografische Information der Deutschen Nationalbibliothek: Die Deutsche Nationalbibliothek verzeichnet diese Publikation in der Deutschen Nationalbibliografie; detaillierte bibliografische Daten sind im Internet über http://dnb.d-nb.de abrufbar.
Alle in diesem Buch genannten Marken und Produktnamen unterliegen warenzeichen-, marken- oder patentrechtlichem Schutz bzw. sind Warenzeichen oder eingetragene Warenzeichen der jeweiligen Inhaber. Die Wiedergabe von Marken, Produktnamen, Gebrauchsnamen, Handelsnamen, Warenbezeichnungen u.s.w. in diesem Werk berechtigt auch ohne besondere Kennzeichnung nicht zu der Annahme, dass solche Namen im Sinne der Warenzeichen- und Markenschutzgesetzgebung als frei zu betrachten wären und daher von jedermann benutzt werden dürften.

Bibliographic information published by the Deutsche Nationalbibliothek: The Deutsche Nationalbibliothek lists this publication in the Deutsche Nationalbibliografie; detailed bibliographic data are available in the Internet at http://dnb.d-nb.de.
Any brand names and product names mentioned in this book are subject to trademark, brand or patent protection and are trademarks or registered trademarks of their respective holders. The use of brand names, product names, common names, trade names, product descriptions etc. even without a particular marking in this work is in no way to be construed to mean that such names may be regarded as unrestricted in respect of trademark and brand protection legislation and could thus be used by anyone.

Coverbild / Cover image: www.ingimage.com

Verlag / Publisher:
LAP LAMBERT Academic Publishing
ist ein Imprint der / is a trademark of
OmniScriptum GmbH & Co. KG
Heinrich-Böcking-Str. 6-8, 66121 Saarbrücken, Deutschland / Germany
Email: info@lap-publishing.com

Herstellung: siehe letzte Seite /
Printed at: see last page
ISBN: 978-3-659-67153-1

Zugl. / Approved by: Jakarta, 2013 - 2014

Contents

Preface

Environmental issues in crowded urban region especially in developing countries have attracted many attentions from environmental scientists. Along with the issues on global world warming and green house effects due to CO_2 elevation and other toxic volatile components as a result of huge industrial activities and recent sophisticated technologies in densely populated regions, thus it makes significant impacts on the equilibrium of ecosystem. In the context of crowded urban region and its dynamics development, rapid population growth and various human characters with different socio-economic backgrounds causing the problems of populated urban becomes more complicated. More issues on "city drowned" due to coastal urban region, soft clay land, large modern high buildings, and ice melting due to global warming generate significant enthusiasm toward urban concern. On account of this matter, this brief review is published to give broader insights to readers who are interested in urban matters related to socio-economic disparities and its spatial segregation.

This book has been dealing with socio-economic and biophysical/technical interactions in urban dynamics life related to uncontrolled population growth divided into six chapters in association with poor management on urban planning especially in Asian developing countries, poor coordination on transportation system, solid waste management, wastewater problems highlighting on water quality deterioration, rapid population growth imbalanced with accommodation facilities, and urban modeling particularly application of geo-information techniques and statistical tools, as well as its hybrids. Nevertheless, the discussions in these chapters in association to urban planning, transportation system and population growth are overlapped and cannot be expressed as individual matter. The chapters of solid waste management and wastewater problems are more focusing on technical matters, while the theme of urban modeling is more emphasized on data processing using specific technical tools and sophisticated software in relevancy with urban matters.

In chapter one, the discussion is more highlighted on urbanization, historical development, geographical condition and regional setting of urban planning especially to big cities in Asian developing countries. The problems related to haphazard land-use, urban expansion, mushroom settlements in unplanned urban area, poor city management, any urban consideration mainly addressing on short term need, limited private sectors to accommodate housing facilities are included as illustrations of what happened in those big cities. In addition, water flood problem and water run-off are often found in urban city due to inefficient city administration and environmental degradation, as well as poor human awareness. At present, several urban regions make a comprehensive plan including efficient management of land use for administration sector, business center, industrial region, housing area nearby transportation link, and recreation area in an integrated urban system, as well as conservation of historical monuments.

In chapter two, transportation matter in urban development is a great threat associated with multi-sector network involving local authorities, government institutions, social organizations, public awareness, physical/technical and financial challenges in order to generate sustainable mega urban life. Most of the improvements on transportation infrastructure in transportation management is mainly focusing on short term needs neglecting any long time consideration that causes additional burdens for the overall urban itself. The problem of traffic congestion is believed to be closely related to limited available roads with the increasing number of vehicles being to ownership, and building new highways or expanding roads only solves the traffic problem for a short time period. Some task efforts have to be taken to overcome traffic problems in densely populated urban region by developing an integrated, reliable, accessible and affordable public transportation system, however, several constraints may resist the transfer development linked to different perspectives among policy makers and limited power of private sectors, as well as public awareness. At the moment, mass rapid transportation is viewed as the most efficient and effective public transportation to overcome transportation problems, however, at some place the project development has faced constraints engaged with different importance among decision makers and financial challenge.

Chapter three discusses about solid waste problem and its impact on public health, natural environment, economy and community. Technical programs and engineering planning on solid waste management implicate with incineration, landfill composting, as well as solid waste dumping and recycling. Lacking of technical skill and knowledge on solid waste management as found in many developing countries may impede the importance and function of solid waste treatment. Solid waste dumping has faced several constraints in terms of low knowledgeable scavengers, limited funding to provide urban solid waste dumping, lacking regulations for efficient solid waste management, non optimal conversion from solid waste to energy, poor decision making on management of solid waste disposal, other sectors concerning with investment, feasible study, profitable business, and role function of stakeholders and policy makers. In addition, solid waste recycling is substantial to reduce waste problems through further solid waste treatment to yield valuable products. Solid waste recycling is defined as a process through which materials previously applied are collected, processes, remanufactured and reused. On other occasion, solid waste recycling has been conducted to yield energy through serial treatments. Plastic recycling industry in local areas shows good prospect on account of sustainable environment, opportunities for better job and economic factor related to saving a lot of money to import more expensive plastics. In some rural areas, huge organic solid waste has been fermented to yield methane biogas for power generation. The organic solid waste is collected from nearby waste disposal transferred from big cities. Solid waste from palm oil industry can be recycled to produce other materials such as pulp, mattress, sound insulator, fiber board, plastic filler. Therefore, an integrated solid waste management is very substantial on account of sustainable environment, working employment, and profitable business.

Chapter four is dealing with water problems with regard to water flood, contaminated river water, ground water over exploitation, poor sanitation, inefficient city administration, poor service quality, and large slum area. Poor water quality and inadequate sanitation may cause significant number of waterborne disease such as diarrhea that hit million people in developing countries annually. The quality of drinking water is the key factor to prevent waterborne disease in reality that water is suitable media for transferring microorganism throughout all countries. Overpopulated urban region and many industrial activities cause fresh water shortage due to ground water over exploitation, inefficient water use and poor sewerage system that it turns to water quality deterioration. Large scale water recycling system for enormous need is burdened by limited funding. To overcome fresh water shortage, some places have made water plant including efficient reservoir system provided with oxygen aerators, expansion of water catchment area to accommodate more rainfall water and renewal of obsolete water infrastructure, as well as land conservation in upstream region and land reclamation after land over exploitation. It is observed that many rivers in densely populated urban are heavily deteriorated due to high population density, poor sanitation, many industrial and commercial activities, and lacking policies on urban sewerage system as well as low human awareness. However, a comprehensive water plant linked to increased access area, capacity building, transparency and accountability, as well as customer orientation has been built in some urban regions to yield clean water with regard to lesser diseases and lower mortality number, to reduce water loss and to create sustainable water supply as well.

In the fifth chapter, rapid population growth has been discussed in broader way and poses a huge threat in densely populated urban that it urges policy makers from multi-sectors to make programs concerning with extensive knowledge and understanding of birth control and family planning. School dropout, under age workers, young beggars, under nourished babies, and wild occupations are negative impacts of economic reason. In some places, urbanization is mostly due to net migration and reclassification generating a characteristic feature, namely, a doughnut phenomenon, which the center is getting emptier and the peripheries getting thicker. Net migration is encountered with the number of incoming people subtracted the number of outgoing people, while reclassification is referred to changes of rural localities to urban localities. Besides the primary economic reason, population growth in certain urban region is induced by various migrations engaged with religion reason, ethnic conflicts, hostility among powerful parties, and other political matters. Due to migration of various ethnics and sub-ethnics, a distribution of various languages and religions as well as cultures reflects a characteristic of the urban city. However, the distribution of cultures is less distinctive compared to that of languages and religions and it is often found as mixed cultures due to social adaptation. Growing urban pressures often causes dramatic rise in poverty, large wild settlements with extremely low living standard, low productivity and unemployment. On the other hand, fast population growth rate in an urban city is triggered by changing of ruling government on a reason of hostility in relationship on the spatial political development strengthened by some other matters such as religion, ethnic, social status and employment. In the sixth chapter, several urban modeling are exposed in general review to reveal that urban modeling is beneficial to sgive insights of urban profile. Combined remote sensing technologies, geographical information system (GIS) and spatial regression are useful to show uncontrolled expansion of urban area and losses of arable land. The model reveals population growth, economics and transportation are the significant factors causing urban sprawl affected by socio-economic background. A CA-Markov model applying remote sensing and GIS on the study of urban dynamics reveals built up zone, agriculture region and fallow land as significant factors on the future design of sustainable urban life. A model applying artificial neural network (ANN) is useful to predict impacts of urban and agriculture expansion on wetland area into three divisions related to unrestricted area, conversion of urban-agricultural expansion into proposed green belt region, and wetland restoration zone. The role of numerical analysis applying Runge-Kutta fourth order on integrated wastewater modeling is also valuable for future directives regarding zone divisions with respect to unrestricted region, restricted wetland area and constructive greenbelt zone. The 4S- model, namely, "survey", "stimulate", "strengthen", and "survey", is a model with respect to human attitude, behavioral change and environmental awareness that is valuable to perform sustainable urban lifestyle. The application of simple statistical approach such as multiple linear modeling in the case study of urban wastewater is worth to show urban wastewater profile. The Pearson Correlation and Post Hoc test are important to reveal any chemical/physical interactions between components in wastewater of interest in relevancy with its regional activities. In addition, the Kolmogorov Smirnov test is useful to examine any normal distribution of a given component in urban wastewater, while the Friedman One-Way Analysis of Variance (ANOVA) is valuable to determine any discrepancy of certain component among groups based on mean rank result.

After all, this book on urban development and its related dynamic matters is not the purpose to replace or to compete with other urban disciplines, but rather a compliment to the topic of interest. The positive values of this book are difficult to be measured, but there is no doubt that it gives contribution to a better understanding and outlook particularly on urban dynamics in fast growing cities. It is a hope that readers may get benefits from this brief review to receive broader insights and to predict future directives on rising urban problems.

Rita Sundari
Research Center of Private Universities Coordination,
13360 Jakarta, Indonesia

1

Urban Planning and its Dynamics Development

Urban planning including city planning is a long time and comprehensive program engaged with better living standard, well established management, public administration and private task with regard to friendly environmental development. According to Nigel Taylor (2007), urban planning is a matter related to strategic development as a mixture of thought, techniques and expertise presented in practical and artistic way (en.m.wikipedia.org /wiki/strategic urban planning). However, in general term urban planning involves with a serial of tasks in physical way including designing and shaping a city in broader scale associated with buildings, streets and public spaces to make urban region more attractive, comfortable, functional and sustainable. Urban planning is a multi-discipline subject engaged with city planning, landscape architecture, architecture science, civil and municipal engineering. On the other hand, poor city management in urban planning along with fast population growth and modernization causes serious problems in public transportation system and solid waste management. Nevertheless, poor city management in urban planning is closely related to transportation, solid waste and wastewater problems due to rapid population growth.

In the case of Dhaka city known as one of the most crowded urban area in recent decade, human settlement in Dhaka city had been started in the 12th century, then the Mughals built as a town for strategic reason in the early 17th century. That Dhaka city has undergone different governments and faced many challenges during centuries, it makes the city mostly exist as unplanned urban area (73% fully unplanned) induced by rapid population growth (over 15 million inhabitants at present) and mushroomed residential (62%) (Ahsanul Kabir and Bruno Parolin, 2010; Sohel Mahmud and Shamsul Hoque, 2006). At present, Dhaka city has only 2% functional road from 52% inaccessible road for motorized vehicles covered about 6.5% of the urban area (Sohel Mahmud and Shamsul Hoque, 2006). As Dhaka city developed through military region in the beginning, business center, trading hub, regional capital, provincial capital and now as the national capital, the city has experienced poor condition in terms of overpopulation, environmental distortion, heavy traffic congestion and inconsistent planning (Ahsanul Kabir and Bruno Parolin, 2010).

As Jakarta city that has experienced rapid population growth within 10 years from 8.2 million (1990) to 23.3 million (2000) due to continuous migration for seeking better income and employment in the city, this municipal region has become a sporadic urban area without strong policies on spatial development and population growth resulting poor condition in transport infrastructure and facilities, as well as accommodation (Soehodho et al., 2005). Closely similar to Jakarta city in terms of socio-economic background as developing urban, Manila city has experienced urbanization in two trends, firstly, urban expansion has been increased in the northern, eastern and southern parts of Manila city leading to loss of urban amenity and while the western part is bordered by the Manila Bay thus resisting urban development in that direction, and secondly, urban expansion has been intensified due to proliferation of informal settlements in the municipal centers as well as the establishment of large trading centers make traffic congestion to worsen and urban life loss its comfort.

On the other side, in nowadays Tokyo city changes its trend with regard to re-centralization in the urban region and aging population mixed with decreasing number of children because the Japanese government has focused on the policy of decentralization since 1970s (Soehodho et al., 2005). At that time several suburban regions were attributed as satellites and mitigations on suburban extension as well as transportation facilities had been strengthened. The trend of re-centralization in Tokyo city is strengthened by rapid land value to decline in the central area during 1990s due to severe recession and aging Japanese population combined with reducing number of children. Moreover, the policy of re-centralization on Tokyo city made the Japanese government strengthen redevelopment projects in the central area by searching more economic investments. Some insights considered that the policy of re-centralization might generate other problems in exhausted rural areas in the near future. The policy change from decentralization to re-centralization in Tokyo city with regard to aging population and reducing number of children has generated several problems in transportation plan in the urban area. While Tokyo city emphasizes on re-centralizing policy in the recent decade, Hiroshima city has different trend in which suburbanization of residential and employment remarkably observed during the last three decades with overall population elevating by 1.18 times from 1975 to 1995 (Soehodho et al., 2005). It is apparent that Hiroshima city does not show high population density, however, the urban region has some tendency of high sub-urbanization with regard to distribution scattering of residential and employment. The tendency of sub-urbanization gives such impact on urban transport system both in the network and mode of transportation yielding less development of mass rapid transport in Hiroshima city compared to Tokyo city, thus it shows the tendency of decentralization in order to create sustainable socio-economic life.

Mumbai as a highly crowded city with higher than 11 million people (2001) was originally a cluster of several islands integrated into a solid land through long term geological processes (Municipal Corporation of Greater Mumbai, 2005). As population growth was likely to increase in the suburban areas during 1950 to 2001 due to faster industrial development in the suburbs rather than that in the city, as a result, it caused economic slowing down in the central area of Mumbai (Fig. 1.1). The Mumbai metropolitan region is setting over a land of 4355 sq. km covering Greater Mumbai, Thane, Kalyan, Navi and Ulhasnagar. In addition, the urban area covers 15 municipal regions, 7 non municipal urban centers and 995 villages (Municipal Corporation of Greater Mumbai, 2005). On the other hand, Dhaka city was built through scattered and unplanned partial urban development without systematic guidance yielding an immense land use with unclear administrative region, residential area, business and commercial centers. Indeed, a partial master plan was provided in the latest 1950s, however, Dhaka city has been developed without appropriate assistance and that plan was not updated in next decades affecting the quality of life, transportation system and environmental condition (Ahsanul Kabir and Bruno Parolin, 2010). Dhaka city has many parks and lakes that make the capital more interesting and attractive, but unfortunately, this municipal region experiences poor city management, low efficiency in administrative matters, and large corruption in service provision for decades leading to urban and environmental degradation. Rapid population growth in Bangladesh and attractive life in Dhaka city as a centre of administration, political, economic, social and cultural matters has strongly triggered urbanization for decades causing social and traffic problems, as well as air and water pollution in

3

Dhaka area (Fig. 1.2). It is expected that in the years after 2010 the rate of population growth increased remarkably in next two decades until 2030 due to rapid economic growth and industrial development. The number of urban population is expected to be roughly doubled as well as the number of rural population in the year of 2030 compared to those populations in the year of 1960. It is roughly estimated that the number of total population in 2030 will be higher four times than that of total population after about 70 years.

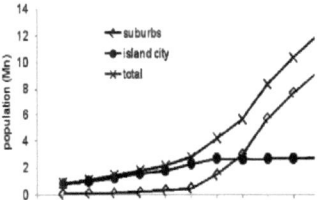

Fig. 1.1 Population trends in Greater Mumbai due to economic factor and industrial development (Municipal Corporation of Greater Mumbai, 2005).

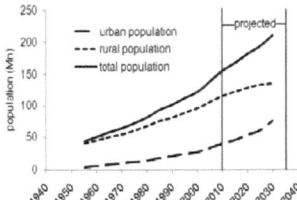

Fig. 1.2 Trend of population growth in Bangladesh with regard to urban regions and rural areas (Ahsanul Kabir and Bruno Parolin, 2010).

The Greater Mumbai area as a whole is a low land, lying on the west side of Sayhidri range. The land is formed by successive layers of basalt flows building step-like terraces and layered appearance. There are five big rivers with numerous tributaries and drain the area into the Arab sea. Moreover, the big rivers play important role as water supply for domestic need of the region. Since Mumbai was built on a basalt land, Jakarta was built on a soft clay land in that 13 river deltas streaming to the Bay of Jakarta. Along with the time, the coastal area of Jakarta has experienced significant land subsidence until sea level susceptible to flooding. According to Kurnia Sari Aziza (2014), land subsidence is observed until 14 cm every year in coastal places and therefore, it is predicted that Jakarta will become one of the drowned cities in the world in 2050. Moreover, if land subsidence is occurred too fast due to land use over exploitation resulting deeper sea water intrusion, it is expected 200 billion USD will be lost and more than 1.5 million people unemployed although a "Giant Sea Wall" project plant along Jakarta coastal area is currently under construction. The Mumbai metropolitan region has a 167 km long coastline with remarkable estuaries, bays and creeks that has panoramic value well known for coastal plantations, beaches and hamlets (Municipal Corporation of Greater Mumbai, 2005). Besides, wetlands are lying on area along the coastal plain consisted of mud, peat and mangrove. Since Greater Mumbai is located along coastal plain with indented creeks and bays, Dhaka is lying on the eastern banks of the Buriganga river and reaches the Ganges Delta that leaves Dhaka susceptible to flooding during wet monsoon due to heavy rainfall and cyclones. In addition, flat region and close to sea level with tropical vegetation and moist soils are the characteristic land of Dhaka causing this urban region undergone heavy flooding annually. As well as Mumbai region with high humidity and heavy rainfall, Dhaka experiences hot, wet and humid tropical climate. While Mumbai area sometimes experiences a pleasant winter, the lowest temperature in Dhaka ever reached 18°C indicated a pleasant mild winter occurred in this region. However, increasing air and water pollution in Dhaka city inducing from overpopulation, traffic congestion and industrial waste make the weather unpleasant affecting human health and its life quality. Moreover, the water bodies and wetlands around Dhaka area are undergone severe destruction as these are worsened by multi-storied constructions and real estate buildings. In coupling with heavy pollution, such destruction of natural habitats seriously influences the life of regional biodiversity.

In terms of poor city management, low management in administration and uncontrollable corruption in public service that is typical characteristics for highly populated urban region in Asian developing countries, as well as Greater Mumbai and Dhaka city, Karachi has developed to a high populated urban area with about 10% of the total Pakistan population but its economic sector is about one-quarter that of its national GDP (Karachi Development Plan, 2000). As Karachi has focused on transportation, storage, telecommunications, wholesale and retail, financial services and real estate as supporting sector, Mumbai city has been dominated by sectors like those sectors of Karachi for supporting its industrial development. For its economic contribution, Karachi has developed some investments in power, chemicals, pharmaceuticals, fertilizer, oil and gas, but Mumbai city has taken attentions on sectors such as gems and jewellery, leather products, information technology and tourism to contribute its economic growth (Municipal Corporation of Greater Mumbai, 2005). Unlike Karachi and Mumbai cities having paid more attention on transportation and real estate to contribute their economic growth, but Dhaka is more concentrated on garment industry to strengthen its industrial sector in that Dhaka exports over 19 billion USD for garments in 2013 (Bangladesh: Jobless rural poor rush to the cities. Integrated regional information networks: UN office for the coordination of humanitarian affairs, 2013). Besides, Dhaka has industrial sectors on jute, cement, ceramics, construction materials and leather goods as well as electronic appliances.

The strategy options to manage housing problems in Mumbai city are oriented on legal land use, affordability to pay for formal housing, minimizing rising cost of formal housing and creating opportunities for investments in real estate (Municipal Corporation of Greater Mumbai, 2005). Most of formal housing in Mumbai city has been built through private sectors, but settlements in slum area are mostly under the responsibility of local government. Fig. 1.3 showed a part of Mumbai city with several new high buildings built in suburb area through private sectors supported by state government as a part of successive transformations. During several decades new satellites have been built to anticipate housing problems in Karachi, however, poor settlers are not interested to stay there due to transportation problems. Therefore, the Karachi master plan takes some considerations by developing new suburbs easily linked to the city center and industrial zone. Another option, the local government of Karachi gives funding supports for people who live in crowded area or re-development the informal settlements with formal housing through private sectors subsidized by the local authorities (Karachi Development Plan, 2000). The willingness of people to live in the congested part of the city is that they may expand their houses as their family grow and they can use parts of the house for running business as well. On the way of people tendency living in more congested regions to expand their houses for running business, Jakarta has similar experienced as well as Karachi. As population growth in densed Dhaka city has escalated so fast, housing problems has come up to surface due to increasing demand for shelters. The ever increasing population from day to day in the 7th largest populous city in the world according to the Bangladesh Bureau of Statistics encourages the government to take some options through clear and sustainable integration of the Mughal settlements in the old Dhaka as a strategic approach to protect Mughal artifacts from decay and to ensure proper access as well as visual exposure in the present urban area (Mohammad Sazzad Hossain, 2013). As back to the past time, Dhaka was established as a provincial city of Bengal in the Mughal period, which is nowadays located in old Dhaka undergone many transformations.

Fig. 1.3 New development in housing sector built in suburb area of Mumbai city as a part of strategic city planning (Municipal Corporation of Greater Mumbai, 2005).

Fig. 1.4 Illustration of a mix land use of "kampongs" and "kota" in Jakarta city due to policies related to housing financing and municipal permit system.

The old Dhaka covers an area of 284 acres with a dense population about 8 million as a home to 15% of total population of urban Dhaka undergone gradual physical deterioration. Like Mumbai city developed new high buildings in suburb area, Jakarta as a metropolitan city has a strategic option in housing sector through vertical development to accommodate high dense population with over 10 million people living in the central area due to limited land space. As Jakarta and its inner peripheries (Tanggerang city, Depok city and Bekasi city) grown to a megacity with total population of around 150,000 in 1900 to about 28 million in 2010 covered an area of around 6,000 sq. km., housing problem is one of the urgent matter that is under the responsibility of the state government to anticipate rapid population growth rate (Hudalah, Delik   and Tommy Firman, 2011). Jakarta experienced some progressive transformations started from the first president, such as a 132 m high national monument namely "Monas" as a symbol of a new tower structure, sport facilities located in "Senayan" used for Asian Games in 1962, "Semanggi" clove bridge, waterfront recreation area located in "Ancol", and "Istiqlal" the big and glorious mosque, as well as new department stores, shopping plazas, new government and parliament buildings and new highways. According to Firman and Tommy (1999), poverty line declines from 40% in 1976 to 11.3% in 1996 that means 6.9 million people in urban district and 15.7 million people in rural area lived under the poverty line in 1996. Gross urbanization from rural area to Jakarta city seeking a better life yielded many villages namely "kampongs" inside the city ("kota"), which is associated with informality, poverty and the retention of rural custom within regional setting. The existence of "kampongs" and modern high buildings reflects spatial segregation and socio-economic disparities. In fact, since 1950 Jakarta has strongly attracted people from all parts of Java as well as other Indonesian islands. Gross migration of people until today is based on the fact that Jakarta has developed to a metropolitan city as centres of politics, administration, economic, social and culture for decades. Although the state government has announced regulations associated with restrictions of new migrants to enter the city, but migrant flooding to Jakarta city for economic reason cannot be avoided as they neglected the law. As a result, the value of houses had continually escalated until two policies were released by the local authorities in relation to subsidies for housing finance program by the regional government and municipal permit system for land development. In fact, these policies have given a lot of benefits to developers closely engaged with the New Order regime and therefore, this new condition generates a capitalist system linked to socio-economic disparities. Half of the land development permits were allocated for 16 development firms, while the other half was distributed amongst the other 167 developers (Michael Leaf, 1994). Fig. 1.4 showed an illustration of a mix land use in Jakarta city comprised of villages and modern high buildings due to socio – economic disparities as a result of housing funding program and municipal permit system for developers.

With regard to the regional setting, the Jakarta city covers an area of about 652 sq. km with population around 10,200,000 comprised of 5 sub-regions, i.e. (i) the northern part of Jakarta covers an area of 142 sq.km with population around 1,190,000; (ii) the southern part of Jakarta covers an area of 146 sq.km with population around 2,080,000; (iii) the western part of Jakarta has an area of 126 sq.km with population around 2,260,000; (iv) the eastern part of Jakarta has an area of 188 sq.km with population around 3,700,000, and the central part of Jakarta has an area of 50 sq.km with population around 962,000 (The Body of Regional Development, 2013). Fig. 1.5 showed the regional setting of Jakarta city in 2014 comprised of 5 sub-regions governed under 5 city-mayors, respectively. The area expansion of northern and central parts of Jakarta city is limited because the northern part of Jakarta is bordered by the Bay of Jakarta. In the reality, the area expansion is prone to the southern and eastern parts of Jakarta in association with industrial area, business centres and housing area. In general the value of housing area in the southern and eastern parts of Jakarta is much higher than the other parts of Jakarta city since these two areas are lying on higher land and the northern part is a flat region with its elevation about the same with sea level susceptible to flooding.

With regard to traffic setting, in some transformations Bangkok city has different policy with Jakarta city follows the similar way as that in Los Angles that the traffic flows in Bangkok city setting in all directions rather than simply to the central core. Due to the rate of urban growth over the past twenty years, the Bangkok metropolitan administration has conduct a study to overcome rising problems in association with traffic congestion, solid waste and wastewater management, and deterioration of air quality as well. As compared to the population in Jakarta city of about 10.2 million occupied a region of 652 sq.km, the population in Bangkok metropolitan city is much more crowded as its population had grown to over than 14.5 million (2010) occupying a land area of around 7.761 sq. km and therefore, the Bangkok metropolitan administration has some visions related to (i) conservation of historical and cultural heritages to maintain its national identity, (ii) preservation valuable national and environmental resources in order to maintain the quality of life, (iii) enhancement of knowledge based on economy through administration and communication centres, (iv) improvement of the city's accessibility by increasing its efficient mass transportation, and (v) development of more efficient land use to accommodate future growth. In order to implement their visions into the real target, the Bangkok comprehensive plan has several proposals to be underlined in relation to (i) polycentric city by setting existing business centre in the central core and community area at the scattered outskirts around, (ii) conservation of cultural heritage in nearby Rattanakosin island, (iii) future public transportation and infrastructure nearby public transportation station by setting up an existing land use, (iv) future development areas to prepare for next coming commercial centres where are easily accessed to transportation stations such as Bansue business centre and Rama III special development zone, (v) extension of urban environment, (vi) ratio in land use of workplaces and residential zones, (vii) major developments in outer ring roads as urban region expansion, and (viii) preservation of rural and nearby agriculture areas in the eastern side to prevent water flooding in Bangkok metropolitan region as well.

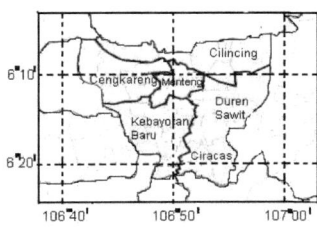

Fig. 1.5 Regional setting of Jakarta city comprised of 5 subregions divided into north, west, south, east and central parts (The Body of Regional Development, 2013).

In general, typical capitals in Asian developing countries are characterized by fast population growth rate, heavy traffic congestion, poor city management, haphazard land use due to inefficient administration and environmental deterioration as well. Those capitals have average population over than 10 million in this decade occupied in imbalanced land area reflecting overpopulation condition.s

Rapid population growth rate is closely associated to limitations in birth control program and as a result, baby booming cannot be controlled. As a whole those urban areas have low elevation of about 4 to 20 meters and are lying on flat regions and therefore, those urban zones experience water inundation. High rainfall and haphazard city planning cause water flooding and inundation. In addition, crowded urban region increases domestic needs on clean water that gives impact on the quality of life. Increasing water intake for long time duration has significant influence on soil structure resulting gradually land subsidence. Several urban municipal regions start their comprehensive plan including management in land use for administration zone, residential area, commercial region and transportation, as well as recreation area and conservation of historical building in the near future. Finally, it is inevitable that urban development needs an integral corporation from many sectors strengthened through environmental management and human awareness.

Transportation in Densely Populated Urban Area

It is inevitable, in fast growing cities with high population transportation matter generates serious problem and traffic congestion, which is often a subject of transportation challenge. Developing urban area is not only faced with solid waste and wastewater problems, but also highly populated municipal area experiences serious traffic congestion and water flooding due to poor city management and haphazard land use. In fact, poor transportation system is a result of administration failure on previous urban development to create a better urban life that is closely associated with multi-sector networks involving local political tenure, government and society organizations, public participation, as well as technical and financial challenges. According to Ortuzar and Willumsen (1994), transportation matter is a very important element in the welfare of a nation and it can be used as a tool to improve viability and contribution to development and welfare. Crises on transportation system have significantly influenced on both physical environment and functional performance of urban development. It is considerably affecting the entire social and physical function generating suffering and inconveniences to people. As a result, the condition will become likely to be out of control unless efficient and effective approaches are taken immediately. With growing increasing displacement due to remarkable growth of urban population and highly occupied settlement, it is of great concern from the local government to generate a sustainable mega urban. However, misunderstanding in factual and root matters causing transportation problems in urban area induced by every uncoordinated approach risen for taking improvements and as a result, it is pushing the urban into a worse situation gradually. Most of the improvements are conducted mainly addressing on short term needs omitting any long term consideration, which often causes additional burden or constraint for the overall urban itself in the infrastructure of transportation development and becomes a great threat for the improvement of transportation management system.

Poor transportation system has often found in highly crowded urban regions especially in developing Asian countries where transportation matters generated after the cities growing too rapidly in accordance with fast population growth. Additionally, many people believe the cause of traffic congestion is closely related to limited roads availability with increase in vehicle ownership. As an example, vehicle ownership was increased by 9 to 11 percent per year, but the growth of roads is lower than 1 percent per year in Jakarta (Asri and Hidayat, 2005). Fig.2.1 presents the graphs of roads development and growth rate of vehicle numbers, which shows inconsistent growth in Jakarta until 2007 (Deden Rukmana, 2010). As quoted from Rukmana report in the Jakarta Post (2010) that developing new roads is not solving traffic problems since building new highways or expanding roads is only reducing traffic congestions for a short period of time. As in the case of Dhaka city known as the 10th most populous city in the world, the building of new roads experiences constraints due to huge populated area (approx. 45,000/km2) with a mushroom development of residential region (62%) and commercial zone (8%) (Sohel Mahmud and Shamsul Hoque, 2006). Because of unplanned development, there are only 1286 km road in Dhaka city including 52% inaccessible roads for motorized vehicles that covered about 6.5% of urban area but in fact, the functional road is only 2% of the municipal region. Since buses are the important mass transportation in Dhaka city occupied only 120 km bus operating route including 22 east-west links, but it is only covered about one-third of the metropolitan area. The possibility to increase roadways and its functionality by implementing low cost traffic management is also faced constraints. As quoted by Saif Asif Khan (2013) from the Asian Development Bank, the number of private motorcycles in Karachi is increased by 9% every year and this adds 280 vehicles every day resulting to massive traffic jams and contributing to high accident rate. Moreover, Karachi has faced daily traffic volumes on major arteries generally in the range of 70,000 - 180,000 vehicles. Theoretically, to overcome traffic congestion in highly populated municipal region some task efforts are taken to develop an integrated, reliable, accessible and affordable public transportation system, however, it is not easily to be implemented in practical way since it faces with a serial constraints linked to local government and private sectors, as well as human awareness. As crowded urban region, the total population of Jakarta metropolitan reached over than 10 million in 2011 occupied a land area of about 300 sq. mile with a variety of sub ethnics from Java (35.16%), Batavia (27.65%), Sundanese (15.27%), Chinese (5.53%), Batak (3.61%), Minang (3.18%), Malay (1.62%) and others (7.98%). In addition, Jakarta has a variety of religions, i.e. Moslems (85.36%), Protestants (7.54%), Catholics (3.15%), Buddha (3.13%). Hindu (0.21%) and

Confucius (0.06%), Besides, the people in Jakarta live with a variety of sub languages, i.e. Indonesian, Batavia, Java, Chinese, Sundanese, Minangkabau and Batak, and also English (Tribunnews, 2014). All the conditions mentioned above take Jakarta as a good reason for a case study of transportation problems related to traffic congestion and water inundation. In the context of transportation in the capital of Indonesia, Jakarta has faced two major problems in the last few years corresponding with traffic congestion and water inundation, which several programs about reducing heavy traffic and flooding in some parts of Jakarta have not properly worked out.

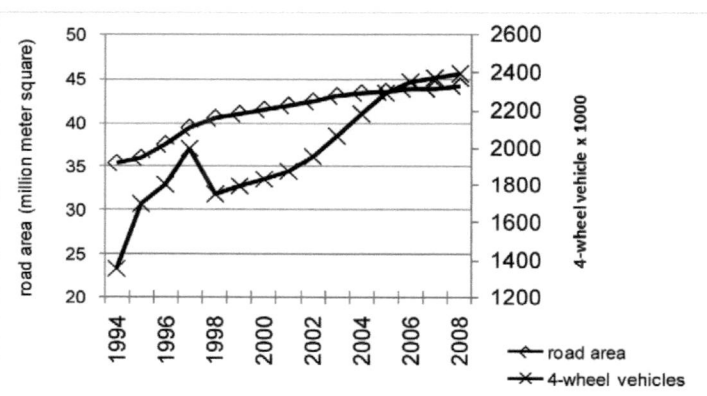

Fig. 2.1 Development of roads compared to vehicles quantity in Jakarta
(Traffic Department, Central of Jakarta, 2008).

This is strong related with the evidence of previous poor city planning failed to generate a better Jakarta for its residents. In order to overcome the traffic constraints the state government in corporation with foreign and local investors made a comprehensive plan to create a better future for Jakarta in terms of meticulous city planning and good transportation system. A poor transportation system may result great lost of valuable expense in relation to poor economic, social and environmental living. In addition, water flood that inundates many parts of Jakarta every year also causes loss of billion dollars. In the context of traffic congestion, building new roads is not the solution to overcome traffic problems since it only decreases traffic congestion for temporary and after a few years, any new road fills with traffic that would not have existed if the road had not been built. Such phenomenon is called as induced demand. As previously mentioned, poor public transportation system is the cause of traffic congestion as existed in Jakarta that its residents tend to use their own vehicles, either cars or motorcycles as their primary transportation because public transportation in Jakarta is not comfortable and insecure and this is the subject of many crowded streets. As reported (Ardiansyah, 2010; Butaru, 2011), the number of modes of public transportation in Jakarta was only about 1.5 percent of the total number of vehicles in 2009. As repeating problems occurred in crowded urban region in many developing Asian countries related to multi-network sectors and role of stakeholders, the Jakarta transport administration failed to realize an integrated, reliable, accessible and affordable public transportation system and therefore, it caused chronic traffic problems.

With regard to public transportation system, mass rapid transportation (MRT) is assumed as the most efficient and effective vehicle to overcome transportation constraints. In the context of MRT's development, as compared to other capitals in Southeast Asian countries like Manila (1984), Singapore (1987), Kuala Lumpur (1995) and Bangkok (2004), Jakarta is too late to build MRT and is expected to finish the MRT project for the first track by 2016 in association with the traffic loop between Hotel Indonesia and the Centrum. The delays of MRT project development in Jakarta is caused by several matters engaged with investment, integral coordination between local authorities and investors, and residential land-free, as well as human factor resisting the development of MRT system in the Indonesian capital. As described by Deden Rukmana in his report (2010), Jakarta was lost about 3 billion USD a year due to MRT delay and will have total traffic gridlock by 2014. Moreover, the delay of MRT building in Jakarta is closely engaged with the history of MRT development in Jakarta. Briefly, the MRT proposal has been discussed by the Indonesian government for at least 20 years and this is the subject to induce corruption among stakeholders engaged with politicians and contractors in relation to 1.6 billion USD loan from the Japanese International Corporation Agency in 2009 (Deden Rukmana, 2010). With regard to MRT system, building commuter lines are assumed to be more effective and efficient rather than increasing number of roads in high urban region since regional trains accommodate a lot of people in one track without facing any traffic congestion on roads. However, Karachi with 13 railway stations has an average rate in the range of 20 – 40 km/hour for working class trains, nevertheless, Karachi is still suffered on economic sector due to loss of time and money caused by densely populated urban area (Saif Asif Khan, 2013). As a brief outlook on Mumbai metro, the first stage operating in June 2014 consisted of 4 – 6 coaches moving along a 7-mile track with average speed of 33 km/h was conducted to overcome transportation problem in severely overpopulated Mumbai city (Wikipedia free encyclopedia). The construction of Mumbai metro consisted of three stages planning over a 15-year period is expected to overall completion by 2021. It is a breakthrough for a highly crowded urban area in South Asian developing countries to build metro transit system in order to reduce transportation constraints and traffic jams.

To generate a comprehensive public transportation system in Jakarta, the MRT system needs to be integrated with other modes of Jakarta transportation system like Trans Jakarta Busway, "metromini" and "kopaja" public minibuses and the city bus. Trans Jakarta Busway was introduced for public transportation in Jakarta in the last decade, but this mode of transportation has faced many challenges linked to poor human awareness contributing to high accident rate since Trans Jakarta Busways took parts in existing public roads and therefore, it is not a solution for high traffic in the crowded urban region as Jakarta city (Noorastuti and Mahaputra, 2009). In addition, a small cab known as "bajaj" accommodated for two or three passengers has been found quite a lot in Jakarta, which the idea of "bajaj" was originally came from public transportation used in India. Fig. 2.2a shows typical public transportation known as "bajaj" that are many used in the Indonesian capital in nowadays. In the history of public transportation in Jakarta, "bajaj" was the idea on replacing older public transportation called as "becak" using human energy by foot pedaling. Another coach for eight to ten passengers is also found a lot in Jakarta carrying people for shorter distance from one to another nearby places known as "mikrolet" (Fig. 2.2b). With a lot of motorcycles in Jakarta and several modes of public transportation, it makes the capital get into poorer condition in public transportation. Fig. 2.3 showed severe traffic congestion during working hours in the central of Jakarta. Nevertheless, a vigorous task has to be taken in the way of simple traffic system by replacing car riders and motorcyclists with a large mass transportation such as MRT. In addition to the development of MRT, Jakarta needs some innovative ways to solve traffic congestion including electronic road pricing, shuttle services, carpool matching services, telecommuting and better parking management in downtown areas. Moreover, several programs need to be implemented in Jakarta related to easy traffic system including developing more districts with a mix of land uses and an interconnected network of roads design to encourage walking and bicycling. While Jakarta is expected to finish its MRT infrastructure by 2016, Manila as one of the capitals in South East Asian developing countries has already operated its metro rail transit system since 1999 (RioHondo, Wikipedia free encyclopedia). However, Manila and Jakarta have similarities in socio-economic characters as crowded urban region in South East Asian developing countries in terms of population growth, centralization, road and public transport investment. Moreover, Manila and Jakarta are still lack of public transport services caused by long development of road infrastructure that accommodate mostly private vehicles. This condition is strongly related to past policies of urban planning that may implicate other sectors corresponding with transport demand in urban areas. As a consequence of long development in transportation networks, the urban planning in developing countries still emphasizes on bus system or other small scale capacity compared to that one in developed countries that emphasizes on higher scale capacity and mass rapid public transport. Table 2.1 shows a comparison of urban public transport in Manila and Jakarta that there is a rather significant discrepancy with regard to traffic congestion between the two cities (Soehodho et al., 2005). It is not surprising that Jakarta as a megacity has problems with traffic congestion since the growth rate of cars and motorcycles from year to year is pretty high as that motorcycles increased faster than cars (Fig. 2.4). Since traffic problem is related to water floods, discussing on water inundation in Jakarta is a substantial topic. Water floods are not a new issue for Jakarta since Jakarta is hit by deadly floods every year and the annual floods inundate many parts of Jakarta resulting a paralyze of the Indonesian's economy. The annual floods cause many people died and urge thousand's of Jakarta's resident's to move away from their homes. A local news reported (Deden Rukmana, 2010) that big annual floods hit the capital in 2010 when the rainfall reached its highest peak causing at least two people died and more than 1,700 people displaced from their old settlements in critical flooding regions in the capital. Moreover, heavy rainfall in recent years caused commuter trains jammed for hours resulting thousands of people trapped in gridlocked streets since the heavy rain paralyzed roads and train traffic. It seems the annual floods occurred in the capital shows that Jakarta has not been able to sustain its urban growth.

(a) (b)

Fig. 2.2 Typical public transportations known as (a) "bajaj" and (b) "mikrolet" are well known in Jakarta.

Back to the pastime where the Dutch colonial government built a system of waterways to protect the capital residents from water floods about two centuries ago, which about 500,000 residents occupied in a lowland region with 43 lakes and 13 rivers relied on the channeling system to prevent water inundation (Zulkifli Hasan, 2013). However, nowadays Jakarta grows to a metropolitan city with its population about 10 millions living within the city and other 4 millions staying in inner peripheries that it still relies on the same waterways system to avert flooding. Besides, water reservoirs, green regions and wetlands that were converted to urbanized areas need to be re-functionalized as non-urbanized areas. Green areas in Jakarta was about 40 to 50% in 1970s, however, the green areas was continually shrinkage until for only 9.3% of its urban area in 2010 (Deden Rukmana, 2010). Therefore, the green area in the capital of Indonesia was reduced to one-fourth until one-fifth after about forty years. With regard to high growth rate of population, inevitably, fast urbanization must be slowed down through distribution of migrants to other prospective and developing areas coordinated by local authorities and government.

Table 2.1. A comparative study of urban transport system of Manila and Jakarta (Soehodho et al., 2005).

City	Population growth	Centralization	Road investment	Public transport investment	Traffic congestion
Manila	+++	++	+	++	+++
Jakarta	+++	+	+	+	+

+ = low; ++ = medium; +++ = high

Fig. 2.3. Severe traffic congestion during peak hours in the central area of Jakarta.

In the reality, water floods give significant impact on transportation and traffic in highly populated urban area. Nevertheless, the migration rate of people from rural areas to big cities must be decreased as one option to reduce flooding problems in lowland municipal region and therefore, it is one solution to solve transportation and traffic jammed in lowland urban region.

On the other hand, Dhaka city as the capital of Bangladesh plays substantial role in economic, business, political and social, as well as culture hub of the country that has turned into one of the busiest metropolitan city with problems related to transportation management system cannot be averted. Since the development process of Dhaka city rapidly increased with tremendous growth of population more than 30 times from 1951 to 2001, Dhaka has undergone proliferation of scattered and unplanned development without systematic approach resulting to transportation weakness and constraints (Habib et al., 2005). Dhaka as the center of major activities in which socioeconomic development and poverty rate of Bangladesh largely depend on Dhaka, it is a great concern to handle sporadic land use and transportation deficiencies. While Karachi in Pakistan as a home of 18 million people compared to lesser population of several countries in Europe like Sweden (9.4 million), Finland (5.4 million) and Norway (4.9 million), it is very interesting to be noticed that a city almost the size of Scandinavia region undergone painful transportation management system in the 21st century. A report said that there are about 24 million people moved in Karachi everyday and about 60% of them dealing with the existing public transportation (Saif Asif Khan, 2013). The total length of roads in Karachi is about 8,000 km compared to Mumbai (1,900 km) and Dhaka (1,850 km), however, this network has to accommodate around 1.81 million vehicles. Table 2.2 illustrates a brief comparison on transportation matters between Karachi and Mumbai in terms of number of vehicle types against population and road length. Mumbai as one of the biggest city in India with about 20 million people living there, more than 88 % of its population uses public transportation including local trains, buses, taxis and rickshaws operating almost 24 hours through the crowded city with a variety of fares in terms of rapidity, convenience and accession that make transportation in Mumbai pretty comfortable (Municipal Corporation of Greater Mumbai, 2005). As many rickshaws found in Mumbai, "jeepneys" are the most popular public transportation in Manila in that its urban area with over 21 million settlements is provided by rapid transit, commuter rail, bus, taxi and "jeepney".

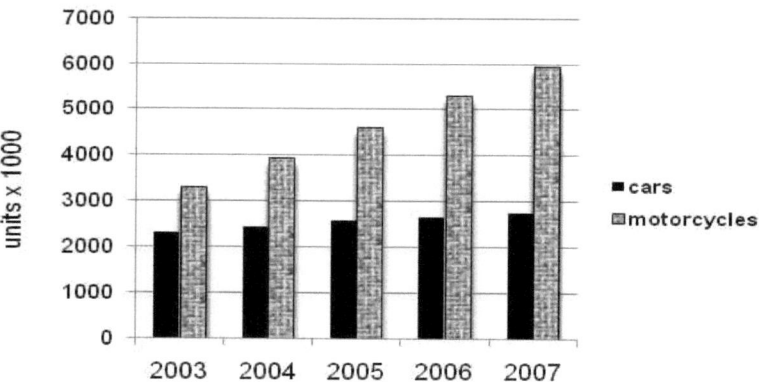

Fig. 2.4 The quantity growth of cars and motorcycles in Jakarta and its satellites (Traffic Department, Central of Jakarta, 2008).

Moreover, migration and settlement of rural residents give many opportunities to get jobs in more developed municipal area. As reported by Betterncourt and West (2010) an expansion of infrastructure of approx. 85% in a municipal region was yielded as the number of population getting about doubled.

Previous reports investigated river basins as favorite places for settlements coming from rural areas because rivers played substantial role in urban development, both socially and economically (Anon, 2003; Shamsuddin et al., 2012). Nevertheless, human behavior from different sub races strengthen by a variety in education background worsened the condition in urban development in relation to waste collection and dumping causing water flooding. In addition, mismanagement in waste collection and waste disposal, as well as poor waterways system causes severe water flooding.

Table 2.2 Transportation conditions in Karachi and Mumbai (Saif Asif Khan, 2013).

	Karachi	Mumbai
Total road length	8,000 km	1,900 km
Population	18,000,000	20,000,000
Number of buses	11,000	25,732
Number of taxis	24,000	98,566
Number of rickshaws	48,000	246,458
Number of citizen per bus	1,500	800

In the case of Dhaka city, urbanization has mainly generated after the independence of Bangladesh in 1971 engaged with centralization of working development, political and administrative power, large investment in developed land, substantial employment chances, disparity in income rate between the city and other regions in the country, therefore, Dhaka city is rapidly developed and faced transportation challenges.

With regard to painful traffic in Jakarta, traffic congestion has been the most transportation challenge as already mentioned above. Therefore, a comprehensive good coordinated and well-managed network involved with multi-sectors including local authorities, private sectors and government is needed to overcome transport constraints. The regional government through private sectors of Jakarta has allocated a budget of about 35 million USD since 2008 in order to reduce traffic burdens (Ardiansyah, 2010). The allocated budget includes public service, vehicles purchasing and controlling management, as well as building roads and city planning. Vigorous task efforts have been implicated considering payment rate, ticketing service obligation, human awareness of using public transportation and simple access to use mass transportation.

Solid Waste Management in Fast Growing Urban Region

Typical highly populated urban region especially in developing countries faces with solid waste problems. Solid waste management has impact on public health, natural environment, economy and society. Moreover, solid waste problem attracts serious environmental issue in association with urbanization, economic growth and living standards particularly in fast growing cities generating a very complicated atmosphere. The environmental factors of fast growth population and urbanization pose a remarkable challenge to environmental scientists to find out solid waste problems. In the context of highly populated urban region, a comprehensive solid waste management is required to solve municipal solid waste problems. Many insights have been described to make programs and planning in linkage with solid waste management getting involved with incineration, waste landfill, and solid waste recycling. Solid waste dumping and burning in open area are common practice in almost all district areas in developing countries. Lacking in technical skill and knowledge in newly developed environmental guidelines and policies, as well as financial capability found in many developing countries may resist the importance and function of solid waste management. However, the solid waste remediation and technology are still under developing process in recent decades. Continual training and capacity building on knowledge and technical skill in solid waste management are intensively required for local peoples who have responsibility for solid waste disposal. Fig. 3.1 showed piles of solid waste in an outskirt of urban region due to untreated solid waste management.

Kuruparan et al. (2011) developed an operation and management of engineered landfill in Sri Lanka by building local capacities through raising awareness and training program for relevant stakeholders and citizens in order to ensure sustainability of operation and management of local facilities. Initial assessments and feasibility studies were carried out to give directions in technical skills, capacity buildings, community awareness and other related issues to solid waste management. Technical guidelines of solid waste management are presented through providing hardware components to local capacities. Modified collection routes and map, as well as additional farm tractors and trailers were provided to increase local facilities. Other matters corresponding with landfill composting were upgraded to improve solid waste dumping. As solid waste dumping being an integral part of municipal solid waste management, Firuza and Nather (2011) designed a solid waste transfer system to accommodate about 2500 tons of municipal solid waste daily in Malaysia through an environmental management plan concerning with air quality, noise, vibration and odor, as well as river ecology, river water quality, socio-economic condition, traffic and transportation issues around the area of interest. Nevertheless, several limitations in practices related to solid waste dumping including low knowledgeable informal scavengers, poor budget availability for providing municipal solid waste dumping, lacking policies and laws for intensive solid waste management, problems on conversion of waste to energy, not well implemented decision making procedures in handling solid waste disposal related to overnight change, and other sectors concerning with opportunities for investment, high profitable business, feasible studies, role of stakeholders including local authorities, organization or government were thoroughly discussed in any occasion such as seminars and conferences in Asian developing countries (Azhar Ali, 2011). In the context of solid waste transfer in highly populated municipal region, Dhaka city with population over 15 million has conducted two stages of solid waste scavenging consisted of primary collection related to initial solid waste selection based on degradability and recycling capability classified as mainly waste from households and the secondary collection in relation to solid waste transfer to disposal sites in each zone (Farhana and Subrata, 2011). That Dhaka city reflected as densely populated urban in developing country is a good case study for the challenging of solid waste management in association with coordination and linkage in solid waste collection and transfer from the root level to community level implicating technical knowledge, human environmental awareness and well set standard cleanliness to build a clean, nuisance and pollution free city. Many environmental events such as seminars and international conferences have been held in Dhaka in last decade with respect to multi sector environmental management. In the case of solid waste recycling that the solid waste is converted to useful material for further material applications is another option in solid waste management concerning with solid waste beneficiation drawing thorough attentions from environmental scientists. As quoted by Schultz in Hossain and Kabir (2011), the term recycling is described as a process through which materials previously used are collected, processed, remanufactured and reused. Besides the solid waste is processed to other benefit substance, solid waste recycling is also an optional way for generating energy through a serial of processes. In the past time, solid waste recycling was conducted mainly due to economic factor since the product was rare or difficult to be produced and therefore expensive, however, at present, the reason is more on ecology consideration since pollutants and residues are viewed to generate more serious problems.

Fig. 3.1 Piles of solid waste in an outskirt of densely populated urban as a result of poor solid waste management.

For instance, plastic recycling industry has shown a mushroom growth in the old parts of Dhaka metropolitan city (Adiba et. al., 2011). The study reported that thousands of workers were employed in more than 250 plastic goods factories in the capital and therefore, plastic recycling was urgent in Bangladesh to sustain environment, create jobs and save million of dollars spent for importing better plastics quality.

In this case, plastic recycling is needed to ensure removal of significant amount of non degradable material and therefore saving the landfill by keeping the quality of soil and environment. Although plastic industry in Bangladesh faced difficulties due to restriction on using polythene bags, nevertheless, the demand on plastic goods was still in high number for export reason and used for packing items. In addition, informal sector plays important role in managing inorganic waste. A study on recycling of inorganic waste in a crowded municipal region showed that residential posed the major source of municipal solid waste (86%), followed by commercial (12%), institutional (1%) (Morshed, 2011). Furthermore, the analysis of municipal solid waste showed plastic as the major generated inorganic waste (62.87%), iron/aluminum/tin (24.69%), glass (11.23%) and battery (1.21%). The related study showed that inorganic waste recycling gave a very profitable business since the business engaged with very low price of waste generation from residential, commercial and institutional sources, good value selling by salvage shops and wholesale, as well as more than 3000 people working in the informal sector related to waste recycling.

Some other illustrations of solid waste recycling to generate other valuable matters are given here. At earlier time, in West Java, Indonesia, a huge organic solid waste was fermented to yield methane gas for electrical power generation of 10.5 megawatt produced by about 20m-height solid waste (Kompas Newspaper, Indonesia, March 22, 2012). The related organic solid waste was collected from waste disposal originated coming from nearby big cities. In some rural areas in Java Island, Indonesia, local people converted solid waste into biogas such as methane for fuel purpose applying simple technology (MetroTV News, Indonesia, September 04, 2014). On other occasion, in East Sumatera, Indonesia, solid waste from animal manure was used as organic fertilizer for palm plantation. On the other hand, solid waste from palm oil industry was processed applying other material for many purposes such as fiber boards, pulp, mattress, sound insulator, plastic filler, etc. (Rosdanelli et al., 2011). Therefore, it is noteworthy to create an integrating system in the process of solid waste recycling to upgrade the benefit of solid waste recycling. Moreover, the smelly solid waste can be advantaged to produce cooking gas in East Java, Indonesia. The process is quite simple, where a tube was filled by solid waste mixed with rice wastewater and allowed for about 7 weeks, the gas produced was streamed out via a plastic pipe connected to a cooking stove. The residual solid waste was used for liquid organic fertilizer (Kompas Newspaper, Indonesia, February 12, 2012). Rosdanelli et al. (2011) in recycling study applied superheated steam drying technology to advantage mass abundance of empty fruit bunches as palm oil waste to generate engineering materials. The production of briquettes from biomass and urban waste is an alternative way of solid waste recycling due to increased fuel price in competition with other briquettes from wood or charcoal. The task was implemented in the Philippine country under integrated cooperation implicating the department of energy, department of environment and natural resources, as well as department of science and technology (Romallosa et al., 2011). The conversion of biomass and urban waste to briquettes gave a positive task to reduce cellulosic waste disposal, to clean the community area from unwanted wastes, to conserve forest and to reduce greenhouse gas effects, and at the same time to provide better livelihood to the urban and poor communities. It is shown here that the solid waste recycling to yield other used material is so important to ensure sustainable solid waste management.

To some extent, solid waste recycling have taken attentions in many developing countries through media, seminars and conferences, as well as programs or courses for extensive training on solid waste treatment. Discussing about solid waste recycling, Faruque and Jahan (2011) proposed an Integrated Solid Waste Management (ISWM) system based on 3R (reduce, reuse and recycle) principle with in relation to poor performance in managing solid waste in highly populated urban area. According to Faruque and Jahan (2011), the small scale, labor level, insufficient technology and unregistered work in informal sectors showed poor service on solid waste management associated with waste recycling activities in terms of waste picking and scavenging. Moreover, the way the solid waste being recycled is very easy in relation to getting high profit and low labor cost where the waste recycling industries give very low payment to waste collectors. In the reality, these waste recycling industries seem to have no any concern towards the local government. In this sense, the environmental awareness of the people has to be stressed and habitual action to achieve successful solid waste management. In addition, the policy makers should play positive role and strengthen regulations/mitigations on waste collectors by considering appropriate revenue, working conditions and social status. As quoted from Getahun et al. (2011), a study on solid waste management was presented in annual occasion by considering quantity, composition, source of waste generated and current disposal practice in growing urban areas in East Africa where solid waste management more concentrates on biodegradable waste applying composting technology since this organic solid waste gave more negative impact on the environment and socio-economic factor. The vision on biodegradable waste is in accordance with the study of Agata et. al. (2011) that biodegradable waste increased with increasing socio economic level since biodegradable waste mostly coming from kitchens and gardens in Ethiopia while glass and plastic materials are reused for recycling purpose.

In broader sense, solid waste management has been discussed in several occasions from many different perspectives. A study on air pollution generation from highly populated urban area in Southern India was conducted concerning with mode of transportation used in that area under study rich with historic temples, monuments, churches and mosques (Horaginamani et al., 2011). As a crowded urban region with population higher than one million, the study emphasized on the way of solid waste transportation collected from different sites using tractors and lorries concerning with narrow streets, dense population, unorganized dumping and poor quality of solid waste. The gas emissions such as CO, NOx and SO2 generated from tractors and lorries using diesel engines were analyzed as pollutant parameters in order to measure air pollution that the NOx emission was the largest among the three gas types under study due to high thermal combustion in diesel engines. Agata et. al. (2011) conducted on specification of solid waste management based on the population income in the capital of Ethiopia in terms of solid waste and biowaste generation. According to the results, the quantity of solid waste was increased as the population increased from middle (10%) to low income (40%) by approx. 50% (Table 3.1). It is inevitable that lower socio economic level consumed more domestic goods since its population quantity is larger than that one of higher socio economic level.

Table 3.1 Solid waste generation in the capital of Ethiopia (Agata et.al., 2011).

	Low income	Middle income	High income
Percentage	40 %	10 %	5 %
Population	1,400,000	350,000	175,000
Waste generated (t/y)	64,792	31,850	14,014
Biowaste generated (t/y)	38,168	21,256	10,099

On the other hand, wastes of glass and metal materials were found in negligible amount since they were not discarded to waste disposal but were used for solid waste recycling. It seems that plastic consumption draws more serious attention rather than other goods consumption as shown by another study conducted on plastic waste management corresponding with income level in populated urban region in Bangladesh (Adiba et. al., 2011). It is inevitable that increased plastic consumption is proportional with increased income generation since higher income level has higher purchasing power (Table 3.2).

Solid waste management in the Visayas region, the Philippines was viewed from the perspective of public awareness and raising support in relation to the role of stakeholders, network building, regional development and ecology center (Paul et. al., 2011). In the report, Paul et. al. (2011) stated that public participation was required for dialogues and consensus in environmental action, as well as forming linkages and networks to enforce the establishment of participatory decision making to make better plan and conduct environmental programs. Moreover, it discussed about the role of local authorities, social organizations, implementation of policies and laws that did not ensure a successful solid waste management due to several limitations corresponding with technical and financial matters. Nevertheless, technical knowledge and budget availability, as well as human awareness and implementation of regional policies are very important to represent successful and sustainable solid waste management as an integral part of non-structural environmental management in highly populated urban area.

Table 3.2   Plastic consumption in Islambagh area, Bangladesh (Adiba et. Al., 2011).

Income level	Plastic consumption (kg/year)	Population (%)
Low	5.24	55
Middle	12.73	40
High	14.01	5

To plan a better solid waste management through common route involving with waste collection, disposal and recovery, an analysis on solid waste from municipal area in Northwest of Iran was carried out corresponding with solid waste chemical composition, moisture and ash contents, density and dimensions (Mohammadi et al., 2011). The results showed food waste had the highest weight (68.9 %), followed by that one of plastic (7.1%) and other components with lower percentages of weight with respect to paper and cardboard, yard waste, textile, glass, and metal (Table 3.3). In addition, the chemical composition of biowaste from the municipal area of interest showed high heat value for methane production as an indication of better energy recovery through methane collection.

Another perspective described that solid waste management is closely engaged with the role of human behavior. In fact, current recycling is facing with several issues and is poorly managed. The role of human behavior in solid waste management depends on human interest in waste recycling engaged with sufficient knowledge and positive attitude towards waste recycling (Hossain and Kabir, 2011). Motivational factors providing with sufficient knowledge and positive attitude are fundamental strategic to make the task of waste recycling simple, significant and interesting. Informal solid waste recycling is usually carried out by poor and marginalized people for income generation and even for everyday survival. According to Hossain and Kabir (2011), depending on where and how material recovery took place, informal waste recycling could be identified in four specifications, i.e. (i) itinerant waste buyers doing waste collection from door to door, (ii) street waste pickers collecting secondary raw materials obtained from waste thrown on streets, (iii) municipal waste collection crew picking secondary raw materials from disposal sites after waste transfer by transporting vehicles and, (iv) waste pickers from dumps in association with communities living in shacks that built from waste construction materials or near the dump.

Table 3.3 The composition of municipal solid waste in Northwest Iran (Mohammadi et. al., 2011)

Component	Wet weight (ton/day)	Weight (%)
Food waste	275.68	68.92
Plastic	28.4	7.1
Paper & cardboard	22.72	5.68
Yard waste	20.44	5.11
Textile	6.2	1.55
Glass	5.04	1.26
Metals	5.12	1.28

To some extent, solid waste recycling is the way to ensure sustainable solid waste management. Solid waste management is also getting involved with the usage of living organisms for solid waste treatment to yield valuable feed for livestock. A pilot study on the use of black soldier fly larvae (Hermetia illucens) was conducted on biowaste treatment to produce a valuable protein for animal feeding and therefore, the biowaste treatment could reduce the mass of organic waste significantly (Stefan et. al., 2011). The fly larvae may process the organic waste into a highly valuable protein source during their last larval stage and that the valuable protein gives a profitable business and may cover parts of waste collection cost. On the one hand, the fly larvae is an extremely resistant species capable of dealing with drought, food shortage or oxygen deficiency in environmental condition. On the other hand, in environmental conditions where the waste source turning into anaerobic, temperatures escalating to lethal values or concentration of heavy metals exceeding threshold level that may affect to larvae population causing fatal condition. However, the business value in the related study in generating innovative and small-scale entrepreneurs is more prospective and may overcome the constraints. In terms of waste composting, the implementation of Vermicomposting technology in solid waste management has been developed as a potential strategy to cover solid waste problems. Roy et. al. (2011) applied a vermicomposting technology to manage organic fly ash from power plants in India using an earthworm (Eudrilus euginae) where the activity of earthworms increasing population of bacteria and fungi through microbial reaction to elevate the degradation rate of organic fly ash. In this study, the organic waste as a mixture of vegetable market waste, cow dung and leaf litter mixed with fresh non weathered fly ash from a thermal power plant was composted with inoculated earthworms for a period of 60 days at 27-33°C. Since fly ash containing nutrients required by crops and giving beneficial impact on the physical and chemical condition of soil supported by organic waste from municipal area this investigation shows a real worthy study in the near future in engagement with vermicomposting method to overcome solid waste problems. In certain region, phenol and its derivatives were detected as major environmental contaminants coming from several industries related to petroleum refineries, steel plants, pulp and pare, pharmaceuticals, synthetic chemicals, coal conversion, etc. Inevitably, phenol is really harmful to human, plants and aquatic life since inhalation, ingestion and skin absorption of phenol cause fatal condition. Therefore, phenol removal using physical, chemical or biological method is of great importance for environmental protection. Biodegradation of phenol in an anaerobic batch reactor using mixed culture in the presence of glucose as substrate was carried out as a pilot study in an industrial region in Iran (Roya et . al., 2011) since phenol removal by biological method is the most favored technique due to low cost, complete degradation of phenol and less remainders. The pilot study was carried out to investigate the rate of biodegradation with respect to various initial phenol concentrations, incubation time and pH monitored during biological experiments. In terms of waste compost, heavy metals like nickel and chromium may contaminate foodstuffs through municipal solid waste used as compost for plant growth. It is known that heavy metals may decrease crop productivity and reduce soil microbial activity,   therefore, a study on light textured alluvial soil contaminated by nickel and chromium using municipal waste compost was investigated in West Bengal, India with respect for nickel and chromium concentrations in soil, spinach leaf biomass from farming field, as well as alternate wetting and drying condition for soil treatment Jayanta et. al., 2011).

4

Urban Water Problems in Populated Region

In a very crowded urban area, clean water is the main problem for domestic need. The term of "clean water" in this context is limited to domestic water for high populated urban area or urban megacities that can be utilized as drinking water after proper water treatment and used for other household purposes as well. Indeed, clean water as the basic need becomes very worthy especially in urban region with high population growth rate on account of standardized water quality and large number of population. Ideally, the quality of drinking water should follow the standards established by WHO (World Health Organization). According to WHO, the quality of drinking water is the key pillar of primary prevention and control of waterborne disease that water can transmit disease in all countries from the poorest to the wealthiest, for instance, the most predominant waterborne disease like diarrhea causes about 2.2 million deaths every year. Moreover, presence of chemical contaminants in drinking water such as naturally occurring arsenic and fluoride for long time exposure is suspected to cause cancer and tooth or skeletal damage. Besides, a lack of proper management of urban and industrial wastewater or agricultural run-off water can substantially elevate the concentration of chemical pollutants in drinking water and long time exposure to polluted drinking water can have serious health implications.

Many high populated urban regions particularly in Asian developing countries have difficulties to obtain clean water that appropriate for domestic purpose. Due to poor city management, many megacities in developing countries have not provided appropriate facilities of domestic water for their citizens. For instance, the soil water in Jakarta is no longer appropriate for domestic water due to intrusion of sea water into the urban land due to ground water depletion as a result of water overexploitation to accommodate overpopulated domestic need and industrial purposes. This matter is very common found in many places in densely populated urban regions located near coastal area. In addition, rapid urbanization has accelerated soil water deterioration that it turns to cause poor water quality. As a result, the soil water has no longer been able to accommodate for domestic need and create water problems. In order to cover the water problem, the state government through social health sector has provided water treatment system for water recycling, however, large scale water recycling is a high cost project required large funding accommodated by the state government. Along with this current water problem, some research institutions have investigated techniques related to water treatment, for instance, distillation method to remove contaminated minerals, ion exchange technique to discard undesired ions, coagulation technique to deposit undesired suspended solids, radiation method applying UV radiation to kill bacteria and other microorganism, ultra membrane for filtration technique, etc., unfortunately, all of those methods for enormous domestic need are non economics. Despite waste water problems in populated urban, on other occasion, wastewater treatment and reuse shows a good prospect on the integrated system of farming and community applying wastewater as a source to yield alternative bio energy. The multi stage-wastewater treatment to yield namely, grey water, includes primary settling process, secondary biological processing via filtering and tertiary process using green algae. The green algae is useful to remove all undesired constituents in wastewater and contains high amount of hydrocarbons as biodiesel energy source, therefore the overall process yields the prospect bio energy grey water (Kumar et al., 2011).

As an illustration, an investigation on urban water addressing to water management   (Kompas newspaper, November 27, 2014) reported that Jakarta as other four cities in Asia Pacific region, namely, Singapore, Tokyo, Shenzhen and Taipei has experienced clean water problems with regard to fresh water shortage mainly caused by sedimentation and groundwater depletion. The decline of surface water in Jakarta is suspected to be the cause of water quality deterioration. In order to overcome this problem, the state government of Jakarta city has made policies on water management in association to efficient water reservoir system, expansion of water catchment area to accommodate more rainfall water, and renewing obsolete infrastructure of water installation. In addition, several task efforts are strengthened to maintain land conservation particularly at the surrounding river areas through regulations engaged with deforestation restriction and land rehabilitation on a high land area of about 18,000 ha at the upstream area. For the importance of investigation, some water samplings were done in Jakarta metropolitan region and other cities in Indonesia by reliable private sectors to examine water quality periodically as the number of population growth elevated.

As well as water problems in Jakarta, Dhaka city faces water problems such as flooding, polluted river water, groundwater depletion, inadequate sanitation, poor administration, haphazard urban development, and large slum area. With regard to water flood, Dhaka city experienced serious flooding for almost four decades (1970 – 2009) and even remarkable flooding up to 4.5 m in some parts of the city in 1988 and 1998. Water flooding in Dhaka city is caused by high rainfall and water overflows from surrounding canals and rivers. Regarding to poor water administration, the Asian Development Bank report (October 2005) on Dhaka water service survey investigated water service for 9 million inhabitants in a 12 million population of Dhaka city, 39% of residential customers received continuous water supply, and only 10% of citizens received drinking water directly from the tap. Due to poor water administration, there is a notable discrepancy between survey results and utility data, for instance, survey's result indicated only 42% of customers received continuous supply but utility data recorded 70% of them received continuous supply, and survey result indicated only 77% of meters in working condition but utility data recorded 97% of them in working condition (Asian Development Bank report on Dhaka water service survey, 2005). On the other side, the water and wastewater management in Metro Manila was found a slow expansion in water access and service quality in the late of 1990s, especially in western region of Metro Manila. Regarding to groundwater problem, Dhaka city has experienced groundwater depletion because of groundwater over exploitation that 577 deep wells has been exploited for 82% city's water need and only the remaining 18% was treated by small water treatment plants. (OOSKA news, 2012). On the other side, Bangkok faces groundwater contamination, this is because many households in Bangkok metropolitan region use indoor flush latrines connected to septic tanks or leaching pits allowing fluid effluents to flow into the ground and collected by a vacuum truck, then disposed at soil treatment plants. Since septic tank has limited capacity, the discharge of effluents creates environmental problem in relation to decline of surface water. Besides, the groundwater surface is high and therefore, the leaching pits are not in proper function. As a result, there is a possibility of flow exchange between groundwater and leaching pit wastewater leading to increased groundwater contamination (Bangkok State of the Environment, 2001).

Due to inadequate sanitation in Dhaka city, there is only one wastewater treatment plant with a capacity of 120,000 cubic meters per day to serve for about 38% of Dhaka population and about 30% of its population uses conventional septic tanks, and another 15% uses bucket and pit latrines as well (Taqsem Khan, 2011). Having concerned with sanitation in Metro Manila, the progress has been far below the contractual target. An increase of access on sewerage system will be expected up to 66% in west Manila and 55% in east Manila by 2021. Regarding the sanitation in Bangkok region, many canals are found to be highly polluted due to direct discharge of wastewater throughout the urban area. It was observed that domestic wastewater was mostly discharged to public drains without further treatment although there were septic tanks to receive toilet wastes, however, the septic tanks generally had outlets to public drains or canals (Thailand Development Research Institute, 1996). In years later, the polluted level of industrial wastewater was getting lesser than those observed in the earlier period, this is because the Thailand government relocated industries to outside area of Bangkok; the main source of wastewater pollution in Bangkok was obviously assumed from the domestic sector (The master plan on sewage sludge treatment/disposal and reclaimed wastewater reuse in Bangkok, 1999). On account of this water pollution, a canal water improvement project was implemented in Bangkok metro region under JICA control in 1990. The project applied re-circulation of water cleaning to the canals and to oxygenate canal water with aerators. This system also used water gates on the canals to prevent salt water entering canals at high tide. Besides, the Bangkok Metro Authority (BMA) in 1990 improved the community wastewater treatment plants and as a result, these treatment plants are in good operational condition and the treated wastewater meets the building effluent standards. To overcome wastewater problems, the Jakarta state government established two integrated Wastewater Treatment Plants or namely, Jakarta Sewerage in 2014 under the cooperation of Ministry of Public Works and Japan International Corporation Agency (JICA) as financial contributor. The Jakarta Sewerage has served five Jakarta sub-regions based on the geographical conditions in each region. At present, the Jakarta Sewerage has implemented 4 projects linked to (i) sewerage planning and construction controlled by Ministry of Public Works, (ii) storm water drainage management under control of public works agency, (iii) environmental and wastewater management under control of environmental agency, and (iv) sludge collecting and treatment service under supervised cleansing agency. Regarding the river pollution, the Jakarta state government through environmental agency estimated river water quality from 2004 to 2009 based on pollutant level

(Fig.4.1). Fig. 4.1 reveals that rivers in Jakarta city were heavily deteriorated during that time (2004 – 2009) due to large population number, inadequate sanitation, many industrial activities, and lack of policies on sewerage system. Besides, continued groundwater over exploitation in Jakarta due to rapid population growth until doubled in next 20 years causes serious land subsidence until the city lower than sea surface about 5 – 6 meter (Brinkman JJ in Kurnia SA, 2014). On the other side, the water supply in Metro Manila strongly depends on Angat Dam, which the dam contributes about 70% of the city's water need and irrigates the farming area of Pampanga and Bulacan states.

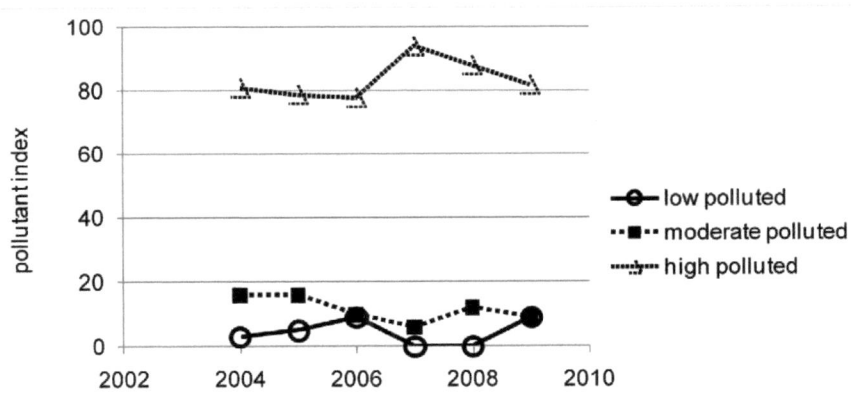

Fig. 4.1 River water quality in Jakarta city region monitoring from 2004 to 2009 (Jakarta Sewerage, Environmental Agency, 2010).

The significant water problem in Metro Manila is the possibility of water shortage in relation to a warning on low water level of Angat's Dam may suffer about two million inhabitants and other six million citizens suffer from weaker water pressure as well. As the water supply depends on Angat Dam water resource, the water shortage in Metro Manila probably can be covered by the El Nino tycoon that often hits the city during November to March every year, however, an analyst (Maynilad and Manila Water news, 2014) reported that the El Nino tycoon had no effect on water shortage in Metro Manila.

In Dhaka city, the low water tariffs cannot cover the operating costs of water utility. According to Dhaka Water Supply and Sewerage Authority (DWASA) management, the low revenues limit the utility's capacity to support higher contribution to investors. On the basis of the figure of water employees per 1000 connections as a common indicator used by DWASA, the water utility in Dhaka city appears to be overstaffed. Based on the result of an observation in 2007, there were 15 employees per 1000 connections, which is almost double that of the average number found in other 18 major Asian cities, i.e. 8.3 employees per 1000 connections in 2001 (Asian Development Bank record). Giving low salaries is the main cause of applying more water employees in Bangladesh rather than that in developed countries, thus, number of staffing should be more concerned. With regard to low water tariffs, the water management in Manila city experienced water crisis because of too low revenues to finance investments to improve water management performance. The water management in Metro Manila has been conducted through water privatization running under two selective water companies, i.e. (i) the Maynilad has operated in west Manila, and (ii) the Manila Water has managed water system in east Manila. However, on the way of running water management, these two water companies has not been able to achieve their targets in terms of increased access area as scheduled in their contracts. Although the Maynilad water company could increase its access, but this water company was unable to reduce water losses and went bankrupt in 2003 due to low water tariffs, financial crisis in East Asia and devaluation in the Filipino currency at that time. On the other hand, the Manila Water company was able to return its contractual rate with hard struggling through the decision of arbitration panel and management strategy on delegation of decision making power to employees at the local level.

The water utility in Dhaka city has implemented a "Turnaround Plan" program (DWASA), which covers capacity building, transparency and accountability, as well as customer orientation. To overcome water problems in Dhaka city, the government of Bangladesh has made a water treatment plant under DWASA since the beginning of 2013 to generate clean water to the whole city yielding lesser diseases and lower the number of mortality in this urban region. The water plant has been funded under corporation between Danish International Development (90%) and the government of Bangladesh (10%). The project has been related to designing and construction of water treatment system that can treat up to 225 million liters of highly polluted surface water every 24 hours. Besides, the water plant will also design a pretreatment water plant with a capacity to operate 450 million liters of water every 24 hours and install a 10 km water supply pipeline with a diameter varied from 600 mm to 1000 mm through very highly populated and heavily trafficked city center. In west Manila, the water supply using pipeline was expanded from 67% (1997) to 86% (2006), customers enjoyed 24 hour water supply increased substantially from 32% (2007) to 71% (2011), and water tariffs was increased significantly by 89% higher than that before 1997. In east Manila, the Manila Water successfully accessed water supply doubled the population number, from 3 million (1997) to 6,1 million (2009). The management of water and wastewater in the eastern zone of Metro Manila is handled by the Manila Water Company in partnership with British and Japanese investors that it deals with water delivery, sewerage and sanitation system, and the Maynilad water company serves to more than six million people in the eastern zone of Metro Manila. In addition, the Manila water company was able to reduce water losses from 67% in 1997 to 11% in 2010, and extended its service area from 26% to 99% for 24-hour water supply. Moreover, the Manila Water company has already spent about 85 million USD to improve the infrastructure of water and wastewater over the last 13 years. Overall, the improvement of water and wastewater management looks faster and more significant in the eastern zone compared to that in the western zone of Metro Manila.

In Bangkok metropolitan region, the water management has been conducted under the Metropolitan Waterworks Authority (MWA), which supplies water averagely 3.8 million cubic meters per day (1991 – 2000). Although the MWA has faced problems of freshwater shortage and deterioration of water quality in the drought period, however, this water council can access 85% of service area for supplying drinking water. Poor water quality in Bangkok metro region is also influenced by improper treatment of wastewater from domestic disposal, agricultural and industrial activities at the upstream of Chao Phraya River. Water supply in Bangkok metro region mainly depends on Chao Phraya and Mae Klong rivers that a fresh water canal was created to convey water from Mae Klong river to MWA's Mahasawat water treatment plant in 2002. Along with promotion of the campaign on efficient water use to maintain fresh water resources and to sustain water supply, the Bangkok state government through MWA has an obligation to maintain the tap water quality in accordance with WHO standard. Therefore, tap water in Bangkok metro region is pretty safe for drinking and it is distributed throughout the city by 13 branch offices. Moreover, the Bangkok state government through MWA has evaluated the impacts of water pollution on three sectors: (i) impact on tourism activities, (ii) impact on aquatic life, and (iii) impact on public health. With regard to the impact on tourism, the deterioration of water quality in many canals located in the inner area of Bangkok gives negative impression on tourists who travel and stay in Bangkok as the capital city and center of tourism. Indeed, high concentrations of toxic sulfide and ammonia or reducing concentration of dissolved oxygen in water can give lethal effects on aquatic life. In the most severe condition that dissolved oxygen concentration in water reduced to zero, all aquatic biota died and this was ever happened in the heavily polluted canals in Bangkok. Even in Chao Phraya River, it was found that some species of fish and plant cannot survive and number of survivors significantly reduced (Bangkok State of the Environment, 2001). Table 4.1 recorded the water quality of Chao Phraya River passing through the central area of Bangkok metro region where the overall water parameters of selected five sampling points are still acceptable. The values of the DO parameter of these selected sampling stations indicate oxygen concentration in river water to ensure the aquatic creatures can survive.

Table 4.1   Water quality of Chao Phraya River in 2000   (Department of Drainage and Sewerage, Bangkok Metro Authority, 2001).

Parameter	Standard B (*)	Nonthabur pier	Rama VI bridge	Chang pier	Memorial bridge	Suphanava pier
BOD mg/l	50	3	3	3	4	3
TSS mg/l	100	69	67	88	69	54
DO mg/l	-	3.9	3.7	3.9	3.4	3.5
pH	5.5 – 9	7.3	7.32	7.34	7.31	7.26
temperature (°C)	40	27.8	27.9	27.9	28.6	28.7

(*) Environmental quality regulations (sewage and industrial effluents), 1979.

According to the Bangkok Metro Authority, a general indication that the water is polluted can be investigated from the following conditions: (i) a substantial reduction of the total number of aquatic creatures that are generally more sensitive than fishes, (ii) a change in the type of species present, and (iii) a change in the number of individuals of each species in the water. Regarding the effects of water pollution on public health, the Bangkok Metro Authority through Disease Control Division reported acute diarrhea was the most expanded disease throughout Bangkok metro region due to inadequate sanitation that it was found 45,000 and 40,000 cases in 1994 and 1995, respectively. However, the number of acute diarrhea cases was even higher in 1997 (Fig. 4.2). In many densely populated regions particularly in Asian developing countries, fresh water resources are gradually reduced because of various water users such as industries, agriculture and domestic need. It is inevitable that the crisis on water needs for coming years is getting more serious. Indeed, the mitigation policies on proper wastewater treatment before discharging into natural receivers should be more strengthened by the authorized bodies along with promotion of the campaign on efficient water needs in the conservation of fresh water resources and sustainable water supply.

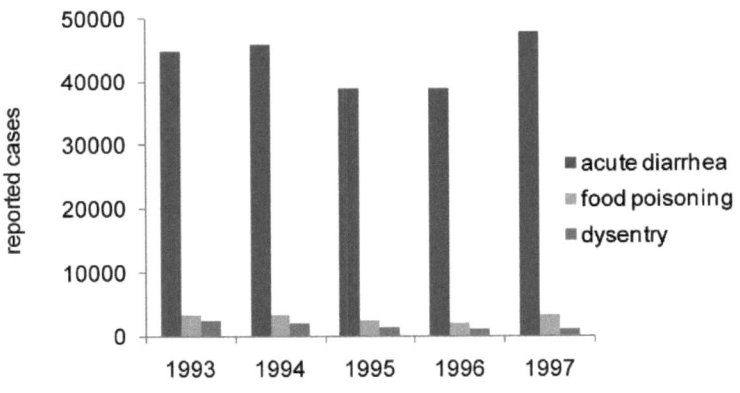

Fig. 4.2 Number of cases of selected water-related disease (Department of Health, Bangkok Metro Authority, 1997).

Furthermore, among the South East Asian developing countries, big cities in Indonesia are still lacking in water infrastructure to get clean water supply compared to crowded cities in Malaysia and Thailand; those cities in the last two countries are more developed in their water infrastructure and architecture including water treatment plants and sewerage system. Compared to Malaysia and Thailand, Indonesia faces another difficulty on account of much larger country land especially for clean water to access. The Indonesian Water Supply Association reported water access over 42% of Indonesian's urban area and 11% of its rural region. In the meantime, most foreigners and middle-income would buy bottled water to drink or water in large containers. Buyers would contribute a percentage from purchasing bottled water to local people, which is a multiple value of what the foreigners do, let say, the Australians pay for drinking water quality. Recently there was a sale of a drinking water quality in Jakarta and there were a lot of consumers from Japan, the Philippines and United States. On the business side, Indonesia has a good prospect for making water supply projects, even that some water projects in Indonesia offered about 18% returns on investment, which would be pretty attractive for investors.

Nevertheless, many densely populated urban regions in some parts of the world have still faced limited access of water supply, poor service quality, lacking treatment of water discharge, groundwater depletion due to water over exploitation, inadequate sanitation, and flooding as well. Enforcement task by the state government through environmental and clean water agency should be strengthened along with the campaign promotion of efficient water use.

5

Population Growth in Rapid Growing Cities

It is known that rapid population growth rate is generally found in many developing countries related to poor family planning, low education and high rate of poverty. In addition, there is a conservative perspective among certain large communities particularly in developing countries that many advantages coming from many children inducing fast population growth resulting in failure of birth control. Besides, high urban migration from rural areas to cities on account of seeking better job and better life has escalated social problems in municipal regions in relation to traffic congestion, water flooding, solid waste and wastewater matters, haphazard land use due to poor city management and inefficient administration, air deterioration, poor human health and environmental degradation, as well as low security insurance. Environmental deterioration found in big cities is getting worse by different socioeconomic backgrounds of migrants claiming enormous demands, consuming more resources and posing various challenges. It is observed that fast population growth generates significant impact on economy, for instance, doubling the population of any city requires about an 85% increase in infrastructure (Battencourt and West, 2010).

For instance, Jakarta city and its peripheries with overall population about 28 million and its population density of around 4500/sq.km has major problems with daily traffic congestion and annually water flooding due to poor management in land use and haphazard city planning. The growth rate of road area is not balanced with the growth of vehicles quantity resulting in transportation problem. In addition, the rate of school dropout is moderately high in recent decade caused by economic impact in this country, although the local government has already given some subsidies in education sector. The amount of the given subsidies from the local government to contribute the economy of low class people is not sufficient enough as periodical funding support. As a result, many school dropout children are going on as beggars or working under age to help their parents at home or to survive for themselves if the parents leaving them. Moreover, overpopulation problem as a result of poor birth control generates many under nourished babies in crowded urban region due to economic factor and lacking knowledge in family planning. According to the national census of 2010, the population   in Jakarta metropolitan region and its nearby satellites has growth rate of 3.6% per annum over the period of 2000 to 2010 (Table 5.1). According to the Central Statistics Agency (2010), the urban area contributed 88.5 % of the total population in the city region or it reached at least 25 million people, however, the population growth in the inner peripheries showed different trend. As shown in Table 5.1, the population growth rates of all inner periphery regions are exceeding the growth rate of Jakarta city, for instance, the population growth rate of Depok city is observed to be 4.30% per annum over ten years.

Table 5.1 Population growth rate in Jakarta metropolitan city and its nearby satellites. (Tommy F, 2011).

Area	Number (million)	2000 – 2010 (%/year)
Jakarta city	9.588	1.40
Bogor city	0.949	2.39
Tanggerang city	1.797	3.12
Bekasi city	2.336	3.48
Depok city	1.736	4.30

It is known that there are three factors in increasing of annual urban population, i.e. (i) the natural population increase, meaning the number of people born subtracted the number of those people died in the same year, (ii) the net migration that is the number of incoming people subtracted the number of outgoing people and (iii) the reclassification referred to changes in rural localities to urban localities. With regard to high population growth rate in the inner peripheries of urban Jakarta region, the trend of 1990 – 2000 was attributed to net migration and reclassification. As a result, the population growth of the Jakarta metropolitan region is referred to as a doughnut phenomenon, which the center is emptier and thicker at the edges and the doughnut is getting bigger and solidifying now. Overall this reflects a spillover of Jakarta metropolitan region with various socio economic backgrounds that it needs a huge scale housing area to accommodate migrant flooding. The following illustrations (Fig. 5.1 to 5.3) are conditions in urban Jakarta region with regard to population growth in Jakarta city over the period 1961 – 2010 (Fig. 5.1), tendency moving to Jakarta city (Fig. 5.2) and population growth of Jakarta and its nearby satellites over the period 2000 – 2010 (Fig. 5.3). Fig. 5.1 shows the increase trend of population growth in Jakarta city mainly due to urban migration. Fig. 5.2 illustrates the percentages of tendency moving to

Jakarta city mostly due to looking for better job and better life, while Fig. 5.3 shows the distribution of population occupied in Jakarta metropolitan region in the last decade (2000 – 2010) attributed as net migration and reclassification. As one of the most populous city in South Asian developing countries, Karachi metropolitan city has a gross population over 23.5 million in 2013 occupied on a land area of 3,527 sq. km, which is known as city's night life meaning that the city never sleeps the whole night.

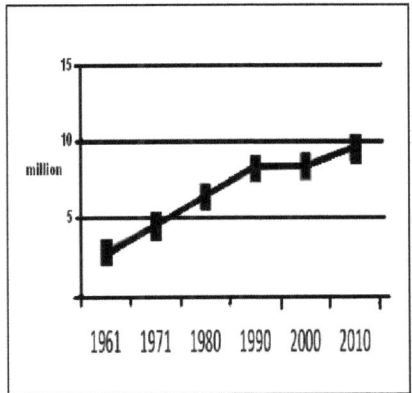

Fig. 5.1 Trend of population growth in the Jakarta city in five decades (Indonesia's urban studies: The megacity of Jakarta, 2011).

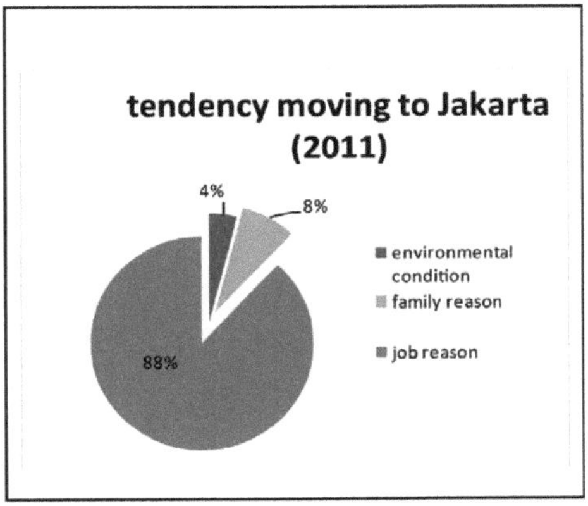

Fig.5.2 Migrants flooding to Jakarta city mostly due to seeking for employment (Indonesia's urban studies: The megacity of Jakarta, 2011).

From the Muslim perspective, Karachi metropolitan is the center of Islamic higher education in South Asia and the largest city in Muslim world, as well as the city of Pakistan's center in banking, industry, economic activity and trading. The population of Karachi city was ever increased about 9 million over 70 years from the beginning 1930s to the latest 1990s (Fig. 5.4), where large scale of migrants came to Karachi metropolitan after Pakistan independence in 1947. This migrants flooding happened when hundreds of thousands of Muslim Muhajirins from India and other parts of South Asia came to settle in Karachi due to religion reason (Geography and Demography, City District Government of Karachi, August 22, 2010). Karachi metropolitan is also known as the city of Quaid, where the Islamic great leader "Muhammad Ali Jinnah" born and died in Karachi. Muhammad Ali Jinnah is recognized as the founder of Pakistan, who made this city his home after Pakistan's independence from the British kingdom on 14 August 1947. Karachi's citizens are composed of ethno-linguistic groups from all parts of Pakistan, as well as migrants from South Asia, making the city with various socio economic backgrounds.

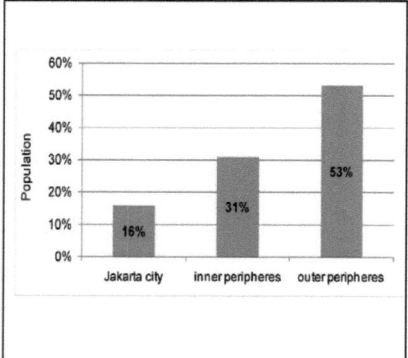

Fig. 5.3 Population distribution in Jakarta metropolitan region over the period 2000 2010 (Indonesia's urban studies: The megacity of Jakarta, 2011).

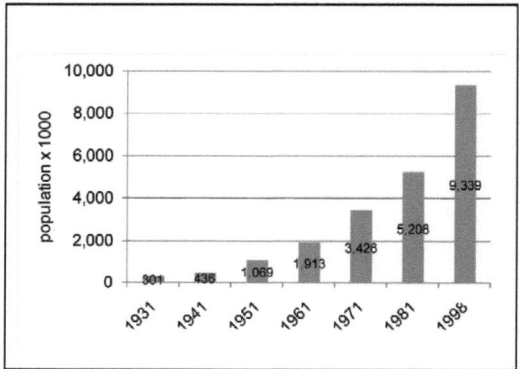

Fig. 5.4 Population growth in Karachi city, Pakistan in seven decades due to urban migration from India and other parts of South Asia (Geography and Demography, City District Government of Karachi, August 22, 2010).

s

In the latest 19th century, the city's population was only around 105,000, with a gradual increase over the next few decades, reaching almost to a million in the year of independence in 1947. Most non-Muslims left Karachi and migrated to India in the 1950s, after the independence, but there are still settlements of Parsis, Goan Catholics and Anglo Indians in the city. After the Pakistan independence in 1947, many Muslim Muhajirs moved to Karachi from India due to anti Muslim programs in India. The culture in Karachi city is a mix culture of South Asia strongly influenced by non-Punjabi Muslim refugees that fled from various Indian states. In nowadays, the descendants of these Muslim refugees are recognized as Muhajirs and become a large and powerful group in Karachi. (Merinews.com., 2012). Besides, Karachi is also a home of 1 – 2 million ethnic Bengalis migrated from Bangladesh in the 1980s to 1990s and followed by Rohingya Muslim refugees from western

Burma. Moreover, Karachi is also a host of migrants from other parts or states in other countries, for examples, many expatriates from Uganda arrived and settled in Karachi as their ancestors known as African slaves, refugees from Iran stayed in Karachi till the late 1980s, settlers from Soviet Union as political or economic migrants, a large numbers of Arabs, Filipinos and Sinhalese from Sri Lanka, expatriates from China working as dentists, chefs and shoemakers, then Polish refugees from Russia migrated to Karachi during World War II, American and British expatriates came and settled in Karachi, Punjabi and Kashmiri Muslims from Punjab and Kashmir valley, respectively, as well as Pashtuns from Khyber, Afghanistan in decades settled in Karachi of about 25% of overall Karachi's population (Karachi Metropolitan Corporation, 2012). With a variety of migrants from different socio economic backgrounds, it forms a linguistic distribution comprised of Urdu (48.52%) as the main language in Karachi, Punjabi (13.94%), Pashto (11.42%), Sindhi (7.22%), Balochi (4.34%), Saraiki (2.11%) and others (12.45%) including Arabic, Farsi and Bengali languages (Karachi population, 2010). Besides, English is spoken language of city's white collar workforce. According to the census of 1998, it also forms a religion distribution in the city with Muslim majority (96.45%), followed by Christian (2.42%), Hindu (0.86%), Ahmadiyya (0.17%) and others (0.10%) including Parsis, Sikhs, Jews and Buddhists (Arif Hasan and Masooma Mohibur, 2010).

As well as Karachi city with Muslims majority, Islam is the predominant religion (83%) in Dhaka metropolitan with population over 14 million, which is mostly belongs to Sunni sect beside small communities of Shiah sect and Ahmadiya society. The substantial development of new trade centers, commerce, industry and administration in Dhaka city has drawn many people from rural areas in Bangladesh moved to Dhaka city in recent decades rising significant economic problems. This situation is reasonable since the real conditions in the development of several megacities, for instance, Shanghai and Guangzhou in China, Delhi and Mumbai in India, Mexico city in Mexico, and Sao Paulo in Brazil, which those cities have experienced substantial migration on a reason of seeking better job and better life (Choong, 2012). Since the infrastructure of Dhaka city is inadequate and unable to keep up the growing urban pressures, this condition generates the dramatic rise in poverty. It is found that most people in Dhaka city are living in slums and squatter settlements with extremely low living standard, low productivity and unemployment. Moreover, the number of the Dhaka city's young population is relatively high due to age selective rural urban migration that about 40% of the total city's population is the unproductive age groups of 0 – 14 and over 60, which shows a high dependency burden on the working age population resulting in poverty particularly among the low income groups (Shahadat Hossain, 2008). The first substantial urbanization in Bangladesh was observed in 1947 where large scale Muslims from India migrated and settled in Bangladesh after Bangladesh had achieved its new political status in 1947. After the partition of India in 1947, Dhaka city became the provincial capital of East Pakistan and the urban population growth started to elevate substantially. In 1951, Dhaka city had a population of only 411,279, which was elevated to 718,766 in 1961. Then there was a significant growth of urban centers followed by an explosive growth of big cities after the independence of Bangladesh in 1971 (Shahadat Hossain, 2008). During the British rule, Calcutta as the primary city of Bengal drew many colonial interests, while Bangladesh became a passive periphery of the region. During the Pakistani rule, hostile relations developed when Bangladesh attempted to become an active periphery of the region. Therefore, the political spatial development process of Bangladesh has passed through passive and active stages followed by urban migration as well as hostile situations. The failure of city planning is directly engaged with urban mismanagement and increasing inequality and poverty in Dhaka city during the period of Bangladesh independency, for instance, the colonial development of a selective exclusive area linked to an urban housing upper class in Dhaka city through the process of suburbanization, which is one of the main causes today's land crisis and the sporadic slums and squatters in Dhaka city (Shahadat Hossain, 2008). Table 5.2 shows the population growth and area extension of Dhaka city during 1931 – 2001 (Population census in Ministry of Dhaka city planning, 2001). On further examination of Table 5.2 with regard to the changing of ruling government from British to Pakistan, there is a sharp increase in the number of population in the beginning of Pakistan period (411,279) about double that under the former British government (239,728), as well as the land area became larger in Dhaka city from 25 to 85 sq.km from 1941 to 1951, while this phenomenon is also observed with regard to the changing of ruling government from Pakistan to Bangladesh associated with remarkable increased number of population in the early Bangladesh period (2,068,353) almost treble that in the previous Pakistan ruler (718,766), as well as the land area grew larger in Dhaka city from 125 to 336 sq.km from 1961 to 1971. The significant increase in population and land area in Dhaka city during the change of ruling government was strongly associated with the hostility of relationship on the spatial political development in Bangladesh strengthened by some reasons such as religion, ethnic, social status, better job and life. With regard to the population growth in India, the population growth rate in Mumbai is twice that of Maharashtra state and 2.5 times that of the country within a century (Ram and Jones, 2013). As Greater Mumbai is known as the most populous city in India, people migration has played important role in the urban development, which 70% of migrants moved to Greater Mumbai coming from Maharashtra state resulting rising problems in Mumbai related to transportation, housing, water and work life balance.

Table 5.2 Population growth and area expansion of Dhaka city in the period of British rule, Pakistani rule and Bangladesh independency (Population census in Ministry of Dhaka city planning, 2001).

Year	Periods	Population	Area (sq.km)
1931	British period	161,922	20
1941	British period	239,728	25
1951	Pakistan period	411,279	85
1961	Pakistan period	718,766	125
1971	Bangladesh period	2,068,353	336
1981	Bangladesh period	3,440,147	510
1991	Bangladesh period	6,887,459	1,353
2001	Bangladesh period	10,712,206	1,530

As a result, Greater Mumbai has induced the growth of satellite cities and out-migration of younger generation who are IT professionals and skilled workers. Besides, Mumbai is the city of dream for many Indians linked to strong Bollywood fascination. Indeed, what is the attractive side of Mumbai? In the past, Mumbai was created during British rule in India beginning its existence in 1661 at the time of the British East India Company occupied a cluster of seven islands and merged together to form as island city, then the island city integrated to Salsette island in the northern part through land reclamation, and today it is known as Greater Mumbai. In order to face many hardships and difficulties, the regional government has already made a comprehensive plan over the development of Greater Mumbai by re-planning industrial, market and office activities in a way that the new urban region will be sustainable from physical, economical and environmental sides. According to Jones W. Gavin. and Mike Douglass (2008), Mumbai has similarities with other four cities in the Asia Pacific region, namely, Jakarta, Bangkok, Manila, and Shanghai, in terms of population growth and land area as mega urban region dynamics that required a systematic study with respect to its various geographic zones, namely, the core, inner zone and outer zones. Table 5.3 shows the core area and population density in the core, inner zone and outer zones of Mumbai, Jakarta, Manila, Bangkok and Shanghai.

Table 5.3 Core area and population density (per sq.km) in different zones of some mega urban regions in Asia Pacific region (Jones W. Gavin and Mike Douglass, 2008).

City and Year	Area (sq.km)	Population density (per sq.km)			
	Core	Core	Inner zone	Outer zones	Total region
Mumbai 2001	603	19,758	8,250	1,071	4,549
Jakarta 2000	662	12,610	3,975	1,085	3,432
Manila 2000	633	15,642	2,047	648	1,641
Bangkok 2000	876	6,709	1,248	472	1,414
Shanghai 2000	605	16,415	1,871	808	2,603

Further examination on Table 5.3, the land area of the core in Mumbai (603 sq.km) is comparable with those in Shanghai (605 sq.km), Manila (633 sq.km), Jakarta (662 sq.km) and Bangkok (876 sq.km). The population density of the core Mumbai is higher than that in other megacities, and the remarkable contrast showed by the Mumbai's inner zone that its population density is by far the highest. The population density in the Mumbai's outer zones is comparable to that in Jakarta, and higher than that in other comparator megacities, where the outer zone is described by Jones (2008) as the remaining area surrounding the inner zone in administrative way defined as the mega urban region (Jones, 2008). Moreover, the population density in Mumbai's entire region is almost 5000 per sq.km as one of the highest population density in the world, almost double that in Shanghai and about treble that in Manila and Bangkok (Jones, 2008; Urban Age, 2007). However, the Mumbai metropolitan region contributes 40% of the GDP of the Maharashtra state as the second most populous state in India with 112 million people in 2011, but only 5% of the national GDP. On the other side, the contribution of Bangkok and Jakarta to their national GDP by far is higher, i.e. 44% and 26%, respectively. This is strong related to the fact that the government structure of mega urban regions does not show their key role in the national economy since this structure is very intricate and multilayered with different administrative procedures in responsibility to urban planning and development (Firman and Tommy, 2003). With regard to issues on economics, Mumbai city ever experienced a substantial economic transition during 1980s and 1990s linked with the closure of textile factories followed by prolonged strikes yielding an enormous scale displacement of engineering, chemical and pharmaceutical industries to outer suburbs. The relocation of industries to outer zones in the 1980s and 1990s was so remarkable that it caused a dramatic decline in the number of population in Mumbai core until lower than 2% per annum during those two decades. The dramatic decline in the rate of population growth in Mumbai core region during these past decades contributed to the generation of a new era in the history of Mumbai metropolitan region, although to some extent there was a roughly unchanging population size in the island city but steadily population growth observed in the suburban areas to the north side. Moreover, the remarkable decline on population growth in Mumbai core region was clearly observed from about 2.3% per annum in the period 1991-2001 to 0.8% per annum in 2001-2011. The spectacular decline observed in population growth rate in Mumbai core region caused substantial shift of industrial structure from the core region to outer suburban areas and significantly affected the trend of migration pattern in the entire Mumbai metropolitan region as well. The shifting of industrial structure from core region to outer zones is assumed to be caused by a number of reasons: (i) government policies on industrial structure and activities linked to environmental and pollution mitigations, (ii) confusing in the regulation of government taxation and other policies, (iii) high costs of basic supplies such as water, electricity and transport, (iv) policies related to labor movements in 1980s, and (v) rising cost of properties in the city (Whitehead, 2008; Sharma, 2010). Previous report said that this trend of migration due to industrial factor has ever occurred in other cities like New York and Tokyo since 1970s on the reason of enormous manufacturing job losses assumed as natural and inevitable process (Sassen, 1993 in Jones, 2008). In addition, the decline in the organized sector and the rising unorganized sector could not overcome the growth of total employment yielding a significant slowed down in the rate of total employment growth (Shaban, 2010) and this has significant effect on population growth and migration trend in Mumbai metropolitan region as well. A huge debate on the future of Mumbai and its planning for urban development has been observed since last decade in that the idea of Mumbai will become a city like Singapore or Shanghai. This idea has attracted many policy makers and elite groups in the corridors of power. However, until today, Mumbai metropolitan region as many other mega urban regions has still faced all these challenges in full measure since the number of population living in areas outside the official metropolitan area is continually increasing. Although issues on population growth and migration for Mumbai are still flared up along with the city planning and governance, nevertheless, the Mumbai city has experienced remarkable deep impacts due to various spatial political development related to significant hostile relations for decades.

In the context of rapid population growth particularly observed in many urban mega cities in Asian developing countries, issues on the future of ideal urban planning and governance in a way that the new urban region will be friendly environmental and sustainable from socio economic side are very appealing to policy makers and elite people in the spatial political development. However, over so many years, any kind of urban development process has been neglected due to inevitable hostile relations among regional government and as a result, the main target of providing affordable facilities even for a very low level of income group appears to be ignored. Nevertheless, several constrains in relation to arrogance of policy makers, confusing in taxation regulations, drawbacks in basic supplies for low income group, high cost of properties in mega cities, and rapid population growth rate are characteristics of high populated urban region that it turns to slowing down the urban development process. After all, an integrated corporation in administrative way on urban metropolitan regions is important to be solidified for city planning and governance especially in many urban regions in developing countries.

6

Application of Modeling for Urban Matter

6.1 General Review

Many environmentalists have attempted to create models on urban matters such as linear regression equation, algorithm simulation applying neural network, factor analysis, non linear elastic model, and some other non parametric statistical examination. Since urban matters are very complicated subject including cases linked to solid waste management, wastewater problems, transportation, population growth and city planning, therefore, many modeling are more highlighted on individual cases. Although many models focus on single case, nevertheless, most urban models do not meet the real picture of urban conditions. Some mathematical rules limiting the application of statistical tools, therefore the statistical equations look "too simple" to be applied in the real situation commonly involved with immeasurable interrelated variables, for instance, in the case study of multiple linear regression model (MLR) of polluted industrial wastewater. In the case of MLR modeling, some water parameters such as BOD, COD, DO, TSS, pH, temperature are not really representative to be used as independent variables giving effects on heavy metal contaminants since there are many other remarkable positive and negative ions in that polluted wastewater.

On a closer examination of urban modeling related to solid waste management and wastewater problems, several modeling exposed in previous reports will be reviewed here as a brief outlook. Interesting studies on landfill cover modeling in solid waste discharge using visual hydrologic evaluation of landfill performance (Agamuthu et.al., 2011), multilayer model as simulation approach for aeration composting process (Bari and Koenig, 2011), qualitative and quantitative estimation of water balance for designing and operating landfills applying hydrologic evaluation of landfill performance (Manandhar et al., 2011), simulation study to investigate the behavior of bottom ash using numerical calculation tool (Le et al., 2011), modeling on integrated wastewater treatment plant comprised of waste stabilization ponds, constructed wetlands and fish ponds applying fourth-order Runge-Kutta equation (Irene et al., 2014), modeling of micro pollutants in wastewater treatment system in terms of occurrence, transport and fate of trace level pharmaceuticals (Laura et al., 2014), modeling of nitrogen removal through a vertical flow constructed wetland using treated domestic wastewater (Ania et al., 2014), and urban drainage modeling to assess wastewater management in developing megacities (Juan et al., 2013).

On the other hand, Choong (2012) more emphasized his model on human behavior and environmental awareness linked to sustainable urban lifestyle and behavioral change through his 4S- model, namely, "Survey", "Stimulate", "Strengthen" and "Sustain" (Fig. 6.1).

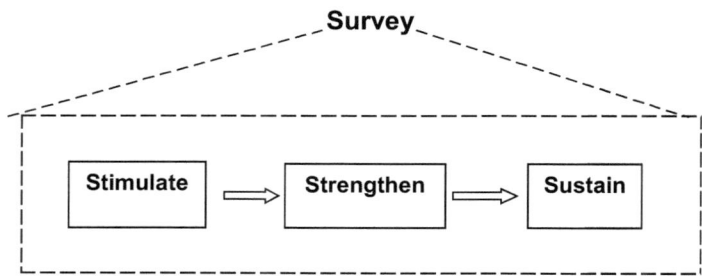

Fig. 6.1 The 4S – model proposed on account of environmental human awareness in urban communities (Choong, 2012).

According to Choong (2012), a comprehensive "survey" including environmental awareness, environmental attitude, as well as environmental practice and concern linked to communal basic needs is important to be implemented with regard to their socioeconomic background. Moreover, the understanding of "stimulate" in his model is associated with the first step in his model to motivate pro-environmental awareness after a comprehensive survey was carried out. For instance, the intentional stimulus is getting involved with educational programs in the way of how to teach people from the earliest stage to love nature or, in other words, how to implement environmental awareness, general knowledge and perception to young children by loving all creatures in nature and its ecosystem. The understanding of "strengthen" in the 4S – model is related to reinforcement between response and stimuli for the purpose of enforcing behavioral improvement. Finally, regarding the last step 'sustain' in the 4S – model, Choong (2012) described that many people are enthusiastic to follow pro-environmental behavior at the beginning only, however, they tend to return to their old habits after certain time. Therefore, the word "sustain" in the 4S – model is required to ensure customers to retain their pro-environmental behavior through repeating actions, both in stimuli and reinforcements.

On the other hand , a SWOT model ("strength", "weakness", "opportunity" and "threat") was developed to profile urban framework encountered to environmental awareness – behavior in densely populated urban region (Rita Sundari, 2014). Fig. 6.2 shows the SWOT model to see urban profile from the view side of substantial factors existed in urban dynamics and development.

STRENGTH	OPPORTUNITY
• Policy maker and stake holder • Environmental scientists (green program and land reclamation) • Expertise on mega city infrastructure and waterways planning • Pro-environmental behavior	• Human activities (better job, better income) • Urban planning (settlement, offices, commercial hub) • Development of industry and technology
THREAT	WEAKNESS
• Water flood disaster and inundation • Land subsidence • Disordered infrastructure and waterways • Traffic congestion	• Low environmental awareness • Increased population • Human attitudes (different socio economic condition, various sub-races) • Illegal buildings and settlements

Fig. 6.2 SWOT model of environmental awareness – environmental behavior related to densely populated urban (Rita Sundari, 2014).

Another insight of urban modeling was developed by a combination of remote sensing, geographical information system (GIS), and spatial regression to assess uncontrolled expansion of urban area and losses of arable land. The results reveal that population growth, economic and transportation is still the main causes of urban sprawl in the way that the expansion of urban built-up land is affected by the socio-economic development at the same period, and significantly influenced by its temporal neighbors (Chen et al., 2015). Shikhar and Akansha (2014) applied remote sensing and GIS through CA-Markov model to study the urban dynamics in rapidly growing cities in order to design a sustainable urban habitat. The results reveal that remarkable changes in 2004 – 2009 occurred in built up classes (27%) followed by agriculture (17.7%) and fallow land (10.2%). The value of kappa (0.91) in the CA-Markov model indicates that the model is valid for future predictions. Vorawit et al. (2014) applied a combining top-view LIDAR data and ground-view SfM observations to create a two-dimensional (2D) urban flood model on the extreme flooding region in a complex urban area. The LIDAR (Light Detection and Ranging) technology was used to provide digital topographic maps and the SfM (Structure from Motion) approach was applied to detect hidden urban features from ground view images, then the combination yielded a Digital Terrain Model (DTM) as the basis of the 2D urban flood model. The results reveal an accurate digital terrain map with respect to floodwater depths and flood propagation patterns around small urban feature. As previous researches focused on urban modeling applying the potential remote sensing technologies and GIS, however, Peter et al. (2014) exposed a critical review on modeling interactions of urban drainage, water supply and broader in an integrated urban water system with respect to numeric considerations such as data issues, model structure, computational and integration-related aspects, common methodology for model development, calibration/optimization and uncertainty. Suggestions of integrated urban water models focused more on interplay between socio-economic and biophysical/technical issues, as well as more user-friendly software application. Chitrini and Nitin (2014) developed a multi-layer neural network based model to predict future impacts of urban and agricultural expansion of wetland area at three different transition rates, namely, simple extrapolation, Markov chain (MC) and system dynamic (SD) applying projected population growth for three different zoning policies. Policies implemented were: (i) "no restriction" for first zoning, (ii) "conversion restriction" of urban-agricultural expansion into the proposed green belt for second zoning, and (iii) "wetland restoration" in the proposed green belt for third zoning. The results reveal that Markov chain was found to be the best match for the projected land demand, while the wetland restoration zoning policies might reduce the impact of population growth on a local scale. The study is valuable for planning and reviewing land use allocation to ensure sustainable wetland ecology. A back propagation algorithm in feed forward artificial neural network (ANN) was developed for linear modeling of oxalate concentration in synthetic wastewater (Rita et al., 2005). A good linear model (R^2 = 0.999) on oxalate determination was obtained with respect to optimization of three factors, namely, learning rate, hidden neurons and number of training cycles. The ANN linear modeling is useful to extend the linear dynamic range of oxalate in urban wastewater. Shaoqing et al. (2014) implemented urban ecosystem modeling (UEM) viewed as inter-disciplinary subject to be used for broader directions in terms of urban ecological properties and their responses to global change. According to this study, the UEM can be classified as top-down models such as materials/energy-oriented models and structure-oriented models, and bottom-up models like land-use oriented models and infrastructure-oriented models, as well as hybrid models. With regard to UEM, a comprehensive UEM approaches at different scales conducts better urban management and more efficient emission mitigation related to environmental pollution and climate change. In order to reduce greenhouse gas effects, a model was proposed with respect to transportation, population density and policies assumed as independent variables on passenger travel related CO_2 emissions (Rabi et al., 2014). The model reveals that adopting policies to reduce emissions in urban area is partly worked by the launching of environmental concerns in that area. Moreover, this study reveals that variables such as population density, transit share, private vehicle and average travel time show significant impacts on passenger travel related CO_2 emissions.

After all, urban modeling has taken major breakthrough and expressed through various ways in the last decade, mostly the application of remote sensing technologies and GIS is exposed at the first place, which is assumed as potential tool to detect urban dynamics based on interactions of socio economic aspects and biophysical/technical issues. In addition, with regard to global issues and climate change, an urban ecosystem modeling has been developed indicating greater concern on environmental pollution due to the development of modern technologies, as well as the application of artificial neural network and numerical analysis software valuable for modeling purpose of urban dynamics in fast population growth.

6.2 Case Study

With regard to the application of statistics tool on wastewater originated from a research institution using many chemicals in its laboratories located at the inner peripheries of an urban area, there are two case studies on research scale discussed in this book, i.e. (i) the implementation of non parametric Kolmogorov-Smirnov (K – S) and Friedman One-Way Analysis of Variance (ANOVA) tests on selected metals distribution of wastewater under study (Rita et al., 2014), and (ii) a multiple linear regression modeling (MLR) modeling provided with Pearson Correlation and Post Hoc test on selected wastewater in urban area (Rita et al., 2013). Both case studies are taken on the same object area, namely, the wastewater from the research institution at the inner peripheries of urban area. Indeed, a stepwise backward MLR using dummy variables was ever implemented on river water model with respect to three zones, namely, upstream area, municipal region, and downstream area, which the model involved with water parameters, selected microorganism (E. coli) and ions (Pb, Cd and nitrate). Based on Wilks' lambda discriminant function, there is a significant difference in terms of predictor(s) between upstream area and municipal region/downstream area at the 0.05 level (Rita et al., 2006). The MLR model reveals that COD was strongly affected by BOD, but less affected by Pb and nitrate.

In nonparametric statistics, the K – S test is commonly used to examine normality of any distribution. If a distribution is proved by the K – S test to be "normal", then the distribution will obey all normality rules, or vice versa. In this context, the K – S was used only to examine whether the metals distribution on wastewater of interest showed a normal distribution or not, and the study is not encountered with any test of metal toxicity or metal pollution. A five selected metals (Cu, Zn, Pb, Mn and Ni) in wastewater of urban research institution taken at three sampling stations on six successive sampling times was examined its normality by the K – S test with respect to each metal concentration. In the K – S test, the lower significance value of a metal distribution reveals the lower degree of normality (Table 6.1). With regard to the scores of 2-tailed significance (Table 6.1), the results show accordingly for Pb (0.896), Mn (0.881), Ni (0.827), Cu (0.293), and Zn (0.108). The result of K – S test in this case study reveals that Zn-distribution posed the lowest normality, or in other hand, the Zn-distribution showed the most decline from normal distribution, while the three other metal distributions (Pb, Mn and Ni) follow the normal pattern. However, this study shows that the

Cu-distribution tends to decline from normal distribution. The normal distribution is valuable for further statistical tests such as estimation, correlation test, ANOVA, and t-test comparison, which it turns to analysis of toxicity and level of water pollution.

Table 6.1 Results of K – S test on five selected metals in ppm (part per million) concentration were taken at three sampling stations on six successive sampling times from urban research institution wastewater (Rita et al., 2014).

Metal	Mean (ppm)	Std.Dev (ppm)	Z-score	2-tail Sig.
Cu	0.09617	0.007771	0.979	0.293
Zn	0.12161	0.109341	1.207	0.108
Pb	0.02661	0.018379	0.575	0.896
Mn	0.03711	0.018381	0.587	0.881
Ni	0.01044	0.006195	0.627	0.827

note: the metal concentrations (ppm) were determined by Atomic Absorption Spectrophotometer (AAS).

With respect to the Regulations of Environmental Quality (1979), the concentrations of selected metals in this case study are still in the acceptable range since the concentrations of selected metals are much lower than those of their respective standard values (Table 6.2). Furthermore, the non parametric Friedman One-Way ANOVA method was used in this case study in order to test the disparity of certain metal among sampling stations of interest on the basis of mean rank value, and the results are shown in Table 6.3. Among those significance values of selected metals, only the significance value of Ni in Table 6.3 shows significant value ($0.011 << 0.050$), this is an indication of disparity of nickel concentration among three selected sampling stations. As one of the three sampling stations under study is a food stall using some corroded cooking utensils, it is suspected as the main subject to cause any disparity on nickel concentration among sampling stations. Both K – S test and Friedman One-Way ANOVA method were processed by SPPS (Statistical Package for Social Science) software under the guidance of George and Mallery (2003).

Table 6.2 Metal data of standard quality and case study.

Metal	Standard (*)	Case study
Cu (ppm)	0.2	0.10
Zn (ppm)	2.0	0.12
Pb (ppm)	0.1	0.03
Mn (ppm)	0.2	0.04
Ni (ppm)	0.2	0.01

(*) Environmental Quality (Sewage and Industrial Effluents) Regulations (1979).

In the second case study, the statistics tool is in dealing with the application of multiple linear regression (MLR) and the process is executed by SPSS software followed the guidance of George and Mallery (2003). The MLR modeling was implemented on the urban research institution wastewater, the same object used for the first case study. However, in this context the MLR modeling implicated with water quality parameters, namely, biological oxygen demand (BOD), chemical oxygen demand (COD), total suspended solid (TSS), pH and temperature besides the selected metal concentration (copper, zinc and lead) taken as variables. As in the first case study, same three sampling stations at six successive sampling times were taken into account for the purpose of MLR modeling. The output data for all variables included in the MLR modeling with respect to three sampling stations at six successive sampling times are presented in Table 6.4. As the lead concentration established as the dependent variable, the general MLR modeling of the urban research institution wastewater is formulated as eq. (1). The toxicity of lead heavy metal is a subject to assume lead as the dependent variable in the MLR modeling. With respect to three sampling stations, three MLR modeling are formulated as eq. (2) to (4).

$$Y_{Pb} = a + bX_{BOD} + cX_{COD} + dX_{TSS} + eX_{pH} + fX_{temp} + gX_{Cu} + hX_{Zn} \qquad (1)$$

That Y_{Pb} is established as the lead dependent variable, and X_{BOD}, X_{COD}, X_{TSS}, X_{pH}, X_{temp}, X_{Cu} and X_{Zn} are independent variables or predictors with respect to BOD, COD, TSS, pH, temperature, copper and zinc. The order of predictors in the general MLR modeling could be changed and depended on the result of regression analysis with respect to its related sampling station.

Table 6.3 Results of Friedman One-Way ANOVA on five selected metals were taken at three sampling stations on six successive sampling times from urban research institution wastewater (Rita et al., 2014).

Metal	Mean rank of sampling sites			Chi-square	df	Sig.
	Station 1	Station 2	Station 3			
Cu	2.50	1.33	2.17	4.333	2	0.115
Zn	2.00	2.00	2.00	0.000	2	1.000
Pb	2.17	1.67	2.17	1.000	2	0.607
Mn	1.83	1.67	2.50	2.333	2	0.311
Ni	1.50	3.00	1.50	9.000	2	0.011

Using the "enter" mode in the MLR processing for Station 1, the results reveal that pH and copper predictors are excluded from the MLR model, and applying the "unstandardized coefficient" on the output, the COD and TSS predictors are omitted from the MLR model, therefore, the MLR modeling for Station 1 can be formulated as follows,

$$Y_{Pb} = -0.236 - 0.05X_{BOD} + 0.016X_{temp} + 0.080X_{Zn} \qquad (2)$$

With regard to the insignificant relationship between COD predictor and Pb independent variable at Station 1, this result shows a consistency with that of previous investigation on MLR modeling of river water through upstream, municipal and downstream area (Rita et al., 2006). Using the similar way as that carried out for Station 1 with respect to "enter" mode and "unstandardized coefficient", the results reveal that COD and pH predictors are excluded from the MLR model for Station 2, and the TSS predictor is omitted from the MLR model, therefore, the MLR modeling for Station 2 can be formulated as follows,

$$Y_{Pb} = -0.676 + 0.001X_{BOD} + 0.012X_{temp} + 3.193X_{Cu} + 0.025X_{Zn} \qquad (3)$$

Table 6.4 SPSS output of urban research institution wastewater with respect to 3 sampling stations at 6 successive sampling times, water quality data (BOD, COD and TSS in mg/L, and temp in °C), and selected metals (Cu, Zn and Pb in ppm concentration); Rita et.al. 2013.

station	time	BOD	COD	TSS	pH	temp	Cu	Zn	Pb
1	1	30.63	673	52	6.59	29.1	0.091	0.053	0.045
1	2	37.50	542	67	6.02	29.6	0.114	0.126	0.042
1	3	38.13	627	89	5.74	28.2	0.092	0.062	0.006
1	4	35.00	593	85	5.87	28.1	0.109	0.126	0.027
1	5	28.13	509	82	6.12	28.4	0.096	0.061	0.067
1	6	31.88	615	92	6.14	28.8	0.110	0.135	0.053
2	1	36.25	633	184	6.12	29.7	0.087	0.045	0.026
2	2	29.38	487	82	5.93	29.3	0.092	0.357	0.022
2	3	30.63	503	91	6.04	28.9	0.095	0.048	0.022
2	4	34.38	437	96	6.01	28.5	0.095	0.348	0.030
2	5	34.38	473	88	6.26	28.7	0.091	0.046	0.011
2	6	30.63	571	124	6.01	28.5	0.088	0.348	0.007
3	1	33.75	374	198	6.09	29.6	0.092	0.090	0.038
3	2	35.63	412	123	6.01	29.4	0.100	0.056	0.024
3	3	32.50	489	109	6.18	28.7	0.090	0.091	0.059
3	4	26.88	614	175	6.11	28.3	0.096	0.052	0.006
3	5	27.50	592	101	6.21	28.1	0.101	0.084	0.033
3	6	30.00	537	94	6.19	28.3	0.092	0.061	0.009

With regard to Station 3, the "enter" mode rejected pH and temperature predictors from the MLR model, while the "unstandardized coefficient" neglected COD and TSS predictors from the model, thus, the MLR modeling for Station 3 can be written as follows,

$$Y_{Pb} = -0.699 + 0.012X_{BOD} + 0.126X_{Cu} + 1.255X_{Zn} \qquad (4)$$

Overall, the MLR modeling of all sampling stations neglected pH variable, it means that the pH has no linear impact on lead concentration in the interest wastewater since the range of pH is very limited (about 6.0 unit value) as shown in Table 6.4. On the other hand, copper and zinc have positive impact on lead concentration, except for Station 1. This is reasonable since those three metals (Pb, Cu and Zn) have similar chemical properties in terms of their solubility in water such as hydroxides, carbonates, chlorides, etc. With regard to the MLR modeling of Station 2 and 3 (eq. 3 and 4), it is found that copper and zinc give impact on lead concentration in the interested wastewater, hence, it is suspected that the corroded copper wires in sampling sites 2 and 3 are the subject of copper contamination in the interested wastewater besides the deteriorated Zn-sacrifice anodes.

In the second case study, the MLR modeling is provided with Pearson Correlation and Post Hoc test. In this context, the Pearson Correlation is exposed only for Station 1. With regard to the results of Pearson Correlation shown at Table 6.5, there is a strong negative correlation between BOD and lead concentration (sign. 0.021) in wastewater Station 1. It is known that Station 1 is surrounded with many food stalls located in the area of research institution. Solid waste of cooked food has induced bacterial growth as a subject to "lead eater". According to previous report, lead contaminant is found in wastewater due to leaching of materials such as plastic, glass tubing, and deteriorated cooking utensils as well (Eneh and Agunwamba, 2011). Besides, a strong negative correlation was found between pH and TSS (sig. 0.044). As the water pH elevated, more flocculants from food residues will form in basic medium of wastewater. In addition, there is a remarkable strong positive correlation between zinc and copper concentrations (sign. 0.001). Similar chemical properties of zinc and copper in relation to transition metals in periodical table, but not lead, are subjected to this interactions. Moreover, the insignificant relationship between COD and BOD for Station 1 shown by the Pearson Correlation at the 0.05 level (Table 6.5) reveals a consistency with the result of previous work on MLR modeling of river water through urban area (Rita et al., 2006). The Post Hoc test reveals that copper concentration between wastewater of Station 1 and that of Station 2 is found to be significantly different. Both test modes are agreed in term of significant difference of copper concentrations between wastewater Station 1 and that of Station 2, namely, the Tukey HSD mode (sign. 0.037) and the LSD mode (sign. 0.015) as shown at Table 6.6. Many food stalls are found in Station 1, while many divisions and departments as well as office buildings with much less food stalls are observed in Station 2. The substantial difference on copper concentrations between the two stations of interest (Stations 1 and 2) is assumed to be relevant with different regional activities between the more crowded and less crowded food stalls regions. Future directives, MLR modeling with its providers looks prospective to predict overall conditions in relation to populated urban dynamics implicated with socio-economic and physical/technical interactions. MLR modeling and its providers is simply to be operated using a friendly-using software like SPSS.

Table 6.5 Output results of Pearson Correlation of urban research institution wastewater at Station 1 (note: encircle values in the table denoted for significant correlation between two variables of interest) (Rita et al., 2013).

		lead (ppm)	BOD (mg/L)	COD (mg/L)	TSS (mg/L)	pH	temp (oC)	copper (ppm)	zinc (ppm)
Pearson Correlation	lead (ppm)	1.000	-.826	-.450	-.201	.598	.354	.149	.042
	BOD (mg/L)	-.826	1.000	.117	.157	-.669	.059	.312	.332
	COD (mg/L)	-.450	.117	1.000	-.257	.347	-.020	-.382	-.202
	TSS (mg/L)	-.201	.157	-.257	1.000	-.748	-.670	.233	.345
	pH	.598	-.669	.347	-.748	1.000	.519	-.280	-.309
	temp (oC)	.354	.059	-.020	-.670	.519	1.000	.333	.206
	copper (ppm)	.149	.312	-.382	.233	-.280	.333	1.000	.969
	zinc (ppm)	.042	.332	-.202	.345	-.309	.206	.969	1.000
Sig. (1-tailed)	lead (ppm)		.021	.185	.351	.105	.245	.389	.468
	BOD (mg/L)	(.021)		.412	.383	.073	.455	.274	.260
	COD (mg/L)	.185	.412		.312	.250	.485	.227	.351
	TSS (mg/L)	.351	.383	.312		.044	.073	.329	.251
	pH	.105	.073	.250	(.044)		.146	.295	.276
	temp (oC)	.245	.455	.485	.073	.146		.259	.347
	copper (ppm)	.389	.274	.227	.329	.295	.259		.001
	zinc (ppm)	.468	.260	.351	.251	.276	.347	(.001)	

For instance, an MLR modeling is proposed to reveal the global profile of a densely populated urban based on simple approach applying only five numerical variables and one string variable with respect to CO_2 concentration in atmosphere as independent variable, and predictors including number of selected vehicles, number of population, amount of tons solid waste, sulfate concentration in urban wastewater on monthly monitoring, and string variable related to regulations/policies in restricted or non restricted urban area. The general MLR model can be written as follows,

$$Y_{CO2} = p + qX1 + rX2 + sX3 + uX4 \qquad (5)$$

Y_{CO2} = CO_2 concentration in atmosphere related to transportation.

X1 = number of specific vehicles related to users indirectly proportional to population growth

X2 = amount of solid waste in tons related to solid waste management.
X3 = sulfate concentration in urban wastewater related to acid rain due to industrial activities.
X4 = uncountable matter related to regulations and policies.

Table 6.6 Results of Post Hoc test for copper variable (ppm) at all stations in the urban research institute wastewater (note: encircled values denoted for significant difference of copper concentration between two stations of interest; sign. ≤ 0.05) (Rita et al.,2013).

Dependent Variable: Copper (ppm)

	Sampling station (A)	Sampling station (B)	Difference (A-B)	Std. Error	Sig.
Tukey HSD	Station 1	Station 2	.011	.004	.037
		Station 3	.007	.004	.215
	Station 2	Station 1	-.011	.004	(.037)
		Station 3	-.004	.004	.595
	Station 3	Station 1	-.007	.004	.215
		Station 2	.004	.004	.595
LSD	Station 1	Station 2	.011	.004	.015
		Station 3	.007	.004	.098
	Station 2	Station 1	-.011	.004	(.015)
		Station 3	-.004	.004	.338
	Station 3	Station 1	-.007	.004	.098
		Station 2	.004	.004	.338

Moreover, the MLR model can be divided based on urban zones, for instance, divided zones according to administration region, housing area, business/commerce centers, industrial region, etc. This MLR model with its providers like Pearson Correlation and Post Hoc test can contribute information about any relationship between respective variables of interest.

Reference

Adiba JE, Sabbir S and Alimullah F, 2011. "Analysis of the recycling of plastic waste: A case study of Islambagh area", Proceeding of Executive Summary of the Waste Safe 2011 – 2nd International Conference on Solid Waste Management in the Developing Countries, Khulna, Bangladesh.

Agamuthu P, Fauziah SH and Khairudin L, 2011. "Landfill cover system modeling for leachate management", Proceeding of Executive Summary of the Waste Safe 2011 – 2nd International Conference on Solid Waste Management in the Developing Countries, Khulna, Bangladesh.

Agata R, Nicolas E and Martin K, 2011. "Residential waste quantification and characterization in Addis Ababa, Ethiopia", Proceeding of Executive Summary of the Waste Safe 2011 – 2nd International Conference on Solid Waste Management in the Developing Countries, Khulna, Bangladesh.

Ahsanul Kabir and Bruno Parolin, 2010, "Planning and development of Dhaka – A story of 400 years", 15th International Planning History Society Conference.

Ania M, Jean MC, Marnik V and Pascal M, 2014. "Modeling nitrogen removal in a vertical flow constructed wetland treating directly domestic wastewater", *Ecological Engineering*, **70**: 379-386.

Anon. 2003. Checking troubled waters. (http://www.ecologyasia.com/news-archives/-2003/feb-03/thestar_20030208_3.htm.).

Ardiansyah F, 2010. "How to reduce traffic jam in Jakarta" (http://wartawarga.gunadarma.ac.id/2010/12/how-to-reduce-traffic-jam-in-jakarta/).

Arif Hasan and Masooma Mohibur, "Urban Slums Reports: The case of Karachi, Pakistan", retrieved 24 August 2010.

Asri DA and Hidayat B, 2005. "Current Transportation Issues in Jakarta and It's Impacts on Environment". Proceeding of the Asia Society for Transmigration Studies, **5**: 1792-1798.

Azhar Ali, 2011. "Facts and myths- modern solid waste management in developing countries", Proceeding of Executive Summary of the Waste Safe 2011 – 2nd International Conference on Solid Waste Management in the Developing Countries, Khulna, Bangladesh.

Bari QH and Koenig A, 2011. "Multilayer mathematical model of forced aeration composting process: A simulation approach for uniform aeration", Proceeding of Executive Summary of the Waste Safe 2011 – 2nd International Conference on Solid Waste Management in the Developing Countries, Khulna, Bangladesh.

Betterncourt L, and West G, 2010. "A unified theory of urban living". *Nature*, **467**(7318): 912-913.

Butaru, 2011. "Vehicles management to reduce traffic congestion in Jakarta". (http://www. scribd.com/doc/58861281/wacana-kemacetan).

Chen Z, Yaolin L, Alfred S and Limin J, 2015. "Characterization and spatial modeling of urban sprawl in the Wuhan metropolitan area, China". *International Journal of Applied Earth Observation and Geoinformation*, **34**: 10-24.

Chitrini M and Nitin KT, 2014. "Geospatial scenario based on modeling of urban and agricultural intrusions in Ramsar wetland Deepor Beel in Northeast India using a multi-layer perception neural network". *International Journal of Applied Earth Observation and Geoinformation*, **32**: 92-104.

Choong Weng Wai, 2012. "The 4S model of fostering pro-environmental behavior among urban communities" in Non Structural Environmental Management, book chapter 8, UTM Press, Malaysia.

Deden Rukmana, Jakarta Post, December 20, 2010.

Eneh OC and Agunwamba JC, 2011. "Managing hazardous wastes in Africa: Recyclability of lead from E-waste materials", *Journal of Applied Science*, **11**: 3215-3220.

Farhana C and Subrata C, 2011. "A study on solid waste management system in Mirpur (zone 7) area of Dhaka city corporation", Proceeding of Executive Summary of the Waste Safe 2011 – 2nd International Conference on Solid Waste Management in the Developing Countries, Khulna, Bangladesh.

Faruque MA and Jahan M, 2011. "A way of incorporating the informal solid waste managers into formal solid waste managers: A case study of Dhaka city", Proceeding of Executive Summary of the Waste Safe 2011 – 2nd International Conference on Solid Waste Management in the Developing Countries, Khulna, Bangladesh.

Firman and Tommy, 1999. "From "global city" to "city of crisis": Jakarta metropolitan region under economic turmoil". *Habitat International*, **23**(4): 447-466.

Firman and Tommy, 2003. "The spatial pattern of population growth in Java, 1990-2000: continuity and change in extended metropolitan region formation", *International Development Planning Review*, **25**: 53-66.

Firuza BM and Nather Khan I, 2011. "Environmental management plan for Shah Alam solid waste transfer station, Malaysia", Proceeding of Executive Summary of the Waste Safe 2011 – 2nd International Conference on Solid Waste Management in the Developing Countries, Khulna, Bangladesh.

George D and Mallery P, 2003. "SPSS for Windows Step by Step. A Simple Guide and Reference". *Update*, 4th ed., Boston: Pearson Education Inc.

Getahun T, Mengistie E, Haddis A, Wassie F, Addis T, Alemayehu E, Dadi D, Gerven TV and Bruggen B, 2011. "Current production and practices of municipal solid waste management in growing urban areas in East Africa", Proceeding of Executive Summary of the Waste Safe 2011 – 2nd International Conference on Solid Waste Management in the Developing Countries, Khulna, Bangladesh.

Habib MA, Ferdous, F and Maniruzzaman KM, 2005. "Examining impacts of transportation on residential property values using geographical information system: A hedonic price model approach", Proceedings of the 40th Annual Conference of Canadian Transportation Research Forum, Canada, p.14-28.

Horaginamani SM, Arif N and Ravichandran M, 2011. "A study on air pollution generation from municipal solid waste transport sector of Tiruchirappalli city, South India", Proceeding of Executive Summary of the Waste Safe 2011 – 2nd International Conference on Solid Waste Management in the Developing Countries, Khulna, Bangladesh.

Hossain MS and Kabir T, 2011. "Recycling the decisive way for sustainable waste management", Proceeding of Executive Summary of the Waste Safe 2011 – 2nd International Conference on Solid Waste Management in the Developing Countries, Khulna, Bangladesh.

Hudalah, Delik and Firman. 2011. "Beyond property: Industrial estates and post-suburban transformation in Jakarta Metropolitan Region". *Cities,* **29**: 40-48.

Irene TA, Yohana L, Senzia M, Mbogo M and Mbwette TSA, 2014. "Modeling of municipal wastewater treatment in a system consisting of waste stabilization ponds, constructed wetlands and fish ponds in Tanzania". *Developments in Environmental Modeling*, **26**: 585-600.

Jayanta KS, Nav RP and Subba AR, 2011. "Protective limits of nickel and chromium for a sensitive agro-ecosystem contaminated through municipal solid waste compost", Proceeding of Executive Summary of the Waste Safe 2011 – 2nd International Conference on Solid Waste Management in the Developing Countries, Khulna, Bangladesh.

Jones, W Gavin, 2008. "Comparative dynamics of the six mega urban regions", in Jones W Gavin and Mike Douglass (Eds. 2008), Singapore, NUS Press, p.41-61

Jones W Gavin and Mike Douglass, 2008. "Urban regions in Pacific Asia: Urban Dynamics in a Global era", Singapore NUS press.

Juan PR, Neil M, Mario DG, Juan PQ and Cedo M, 2013. "Monitoring and modeling to support wastewater system management in developing mega-cities". *Science of The Total Environment*, **445-446**: 79-93.

Karachi Metropolitan Corporation, "Administrator Office", retrieved February 28, 2012.

Karachi population to hit 27.5 million in 2020", *Dawn*, retrieved August 24, 2010.

Kumar RP, Palani S, Selvaray R and Regupathy I, 2011. "Treatment of grey water using hydrocarbon producing green algae *Botryococcus braunii*", Proceeding of Executive Summary of the Waste Safe 2011 – 2nd International Conference on Solid Waste Management in the Developing Countries, Khulna, Bangladesh.

Kurnia Sari Aziza, Kompas news, July 08, 2014.

Kuruparan P, Forouhar A, Sacramento A and Subasinghe G, 2011. "Dumpsite to engineered landfill experience of the Ampara district, Eastern Sri Lanka "builiding local capacities in the operation and management of engineered landfill", Proceeding of Executive Summary of the Waste Safe 2011 – 2nd International Conference on Solid Waste Management in the Developing Countries, Khulna, Bangladesh.

Laura JP Snip, Xavier FA, Benedek GP, Ulf J and Krist VG, 2014. Modelling the occurrence, transport and fate of pharmaceuticals in wastewater systems", *Environmental Modeling & Software*, **62**: 112-127.

Le NH, Abriak NE, Binetruy C and Benerzour M, 2011. "The study of behavior of bottom ash under triaxial stress", Proceeding of Executive Summary of the Waste Safe 2011 – 2nd International Conference on Solid Waste Management in the Developing Countries, Khulna, Bangladesh.

Manandhar DR, Hogland W, Krishnamurthy V and Khanal SN, 2011. "Use of HELP model for estimation of leachate from a Pilot Scale Lysimeter", Proceeding of Executive Summary of the Waste Safe 2011 – 2nd International Conference on Solid Waste Management in the Developing Countries, Khulna, Bangladesh.

Maynilad and Manila Water news, The National Water Resources Board, 26 August, 2014.

Michael Leaf, 1994. "The suburbanization of Jakarta: A concurrence of economics and ideology. Third World Planning", *Review*, **16**(4): 341-356.

Mohammad Sazzad Hossain, 2013. "Strategies to integrate the Mughal settlements in old Dhaka", *Frontier of Architectural Research*, **2**(4): 420-434.

Mohammadi A, Amin MM and Hasanpour M, 2011. "Survey on energy recovery potential of municipal solid waste in Northwest of Iran", Proceeding of Executive Summary of the Waste Safe 2011 – 2nd International Conference on Solid Waste Management in the Developing Countries, Khulna, Bangladesh.

Morshed Nahid, 2011. "A study on the recycling of inorganic waste of Khulna city", Proceeding of Executive Summary of the Waste Safe 2011 – 2nd International Conference on Solid Waste Management in the Developing Countries, Khulna, Bangladesh.

Nigel Taylor, 2007. "Urban Planning Theory since 1945". SAGE Publish. London.

Noorastuti PT and Mahaputra SA, 2009. "Busway is the most convenient public transportation".
(http://us.metro.vivanews.com/news/read/67626_hanya_busway_angkutan_paling_manusi awi_).

Ortuzar JD and Willumsen LG, 1994. "Modeling Transport", 4th ed., Library of Congress Cataloguing in Publication Data.

OOSKA news, 2012. "Bangladesh National Economic Council report".

Paul JG, Fernandez Ricana MV and Acosta VL, 2011. "Emerging regional ecology centers in the Philippines: experiences, benefits, and lessons learned from the Visayas region", Proceeding of Executive Summary of the Waste Safe 2011 – 2nd International Conference on Solid Waste Management in the Developing Countries, Khulna, Bangladesh.

Peter MB, Wolfgang R, Peter SM, David TM and Ana D, 2014. "A critical review of integrated urban water modeling – urban drainage and beyond". *Environmental Modeling & Software*, **54**: 88-107.

Political and ethnic battles turn Karachi into Beirut of South Asia Crescent", Merinews.com., Retrieved November 24, 2012.

Rabi GM, Prem KG, Ashley MW and Andrew JL, 2014. "Modeling the relationships among urban passenger travel carbon dioxide emissions, transportation demand and supply, population density, and proxy policy variables". *Transportation Research Part D: Transport and Environment*, **33**: 146-154.

Ram B Bhagat and Jones W Gavin, 2013. "Population change and migration in Mumbai metropolitan region: Implications for planning and governance", Asia research institute, working paper series no.201, National University of Singapore.

Rita S, Musa A and Lee YH, 2005. "A linear modeling by feed forward artificial neural network for oxalate quantitative determination in synthetic wastewater using Al(III)-ECR complex", Proceeding in the 6th Joint Seminar on Chemistry UKM - ITB, Bali, Indonesia.

Rita S, Musa A and Lee YH, 2006. "Equation modeling of Sembulan river, Sabah, as a case study using backward stepwise multiple linear regression ". *Sains Malaysiana*, **35**(2): 1-7.

Rita S, Tony H, Rubiyatno, Fakhri AM and Madzlan A, 2013. "Multiple linear regression (MLR) modeling of wastewater in urban region of Southern Malaysia". *Journal of Sustainability Science and Management*, **8**(1): 93-102.

Rita S, Rosdanelli H and Fakhri AM, 2014. "Metals distribution (Cu, Zn, Pb, Mn and Ni) in campus wastewater: K-S test and Friedman ANOVA". *Advanced Materials Research*, **864-867**: 1755-1758.

Rita Sundari, 2014. "Environmental awareness and its related factors in highly populated urban area: A short review". International Journal of Plant, Animal and Environmental Sciences, 4(2): 172-177.

Romallosa ARD, Hornada KJC and Paul JG, 2011. "Production of briquettes from biomass and urban wastes using a household briquette molder", Proceeding of Executive Summary of the Waste Safe 2011 – 2nd International Conference on Solid Waste Management in the Developing Countries, Khulna, Bangladesh.

Rosdanelli H, Rita S, Madzlan A, Wan Ramli WD and Rahim Y, 2011. "The beneficiation of empty fruit bunches biomass as palm oil waste by superheated steam drying", Proceeding of Executive Summary of the Waste Safe 2011 – 2nd International Conference on Solid Waste Management in the Developing Countries, Khulna, Bangladesh.

Roy G, Iftikar W and Chattopadhyay GN, 2011. "Vermicomposting biotechnology as an effective solid waste management strategy for recycling fly ash from thermal power plants", Proceeding of Executive Summary of the Waste Safe 2011 – 2nd International Conference on Solid Waste Management in the Developing Countries, Khulna, Bangladesh.

Roya P, Ghasem DN, Nafise M, Zeinab B and Tahere T, 2011. "Biodegradation of phenol in an anaerobic batch reactor using mixed culture in presence of supplementary substrate", Proceeding of Executive Summary of the Waste Safe 2011 – 2nd International Conference on Solid Waste Management in the Developing Countries, Khulna, Bangladesh.

Saif Asif Khan, Monthly Economic Review, July 2013.

Shaban A, 2010. "Mumbai: Political economy of crime and space". New Delhi: Orient Blackswan Private Ltd., p.52.

Shahadat Hossain, 2008. Bangladesh e-Journal of Sociology, 5(1).

Shamsuddin S, Latip NSA, Sulaiman AB, 2012. "Towards sustainable development of the urban waterfront in Kuala Lumpur: The impacts of key players and policies" in Non Structural Environmental Management (Choong Weng Wai), chapter 3, p. 23-59.

Sharma RN, 2010. "Mega transformation of Mumbai: Deepening enclave urbanism". Sociological Bulletin, 59(1): 69-91.

Shaoqing Chen, Bin Chen, Brian D Fath, 2014. "Urban ecosystem modeling and global change: Potential for rational urban management and emissions mitigation". Environmental Pollution, 190: 139-149.

Shikhar Deep and Akansha Saklani, 2014. "Urban sprawl modeling using cellular automata". The Egyptian Journal of Remote Sensing and Space Science, 17(2): 179-187.

Soehodho S, Fujiwara A, Hyodo T, Montalbo C, 2005. "Transportation issues and future condition in Tokyo, Jakarta, Manila and Hiroshima", Proceedings of the Eastern Asia Society for Transportation Studies, **5**: 2391-2404.

Sohel Mahmud and Shamsul Hoque, 2006, "Unplanned development and transportation problems in Dhaka city", International Conference on Road Safety in Developing Countries.

Stefan D, Zurbrugg C, Gutierrez FR, Nguyen DH, Morel A, Koottatep T and Tockner K, 2011. "Black soldier fly larvae for organic waste treatment – prospects and constraints", Proceeding of Executive Summary of the Waste Safe 2011 – 2nd International Conference on Solid Waste Management in the Developing Countries, Khulna, Bangladesh.

Taqsem Khan, 2011. "The performance challenges of DWASA in Global Water Intelligence: Focusing on performance, Global Water Summit", p.50-52.

Tommy F, Jakarta Post, March 26, 2011.

Urban Age, 2007. "Urban India: understanding the maximum city", Cities Program, London: London School of Economics and Political Science.

Vorawit M, Zoran V, Arthur EM and Ahmad FA, 2014. "Urban flood modeling combining top-view LIDAR data with ground-view SfM observations". *Advances in Water Resources* (article in press).

Whitehead Judy, 2008. "Rent gaps, revanchism and regimes of accumulation in Mumbai". *Anthropologica*, **50**(20): 269-282.

Zulkifli Hasan, Indonesia's Urban Studies, May 7, 2013.

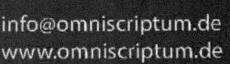